MOTORS AS GENERATORS FOR MICRO-HYDRO POWER

Nigel Smith

INTERME ICATIONS

Intermediate Technology Publications Ltd
103-105 Southampton Row
London WC1B 4HH, UK.

Whilst the author and publishers have made every effort to ensure that the information and guidance given in this work is correct, all parties must rely upon their own skill and judgment when making use of it. Neither the author nor the publishers assume any liability to anyone for any loss or damage caused by any error or omission in the work, whether such error or omission is the result of negligence or any other cause. Any and all such liability is disclaimed.

Reprinted 1997

ISBN 1 85339 286 3

Cover photo: An induction generator system installed as part of IT Sri Lanka's Village Hydro Programme.

Printed in the UK by the Russell Press Ltd

CONTENTS

PREFACE

Micro-hydro is a valuable source of energy for rural industries and village electrification schemes. It has been a traditional method of grain processing throughout the world and played a major role in modernization and industrial development in Europe and North America. Micro-hydro now offers similar potential to most developing countries, with applications in village lighting, mechanized food processing, and the supply of power to small-scale industrial activities.

This book contributes an important element of knowledge which is needed to realize this great potential. In our efforts over the years to assist local manufacturers and installers of hydro equipment, the staff of ITDG have found that conventional generators are difficult to obtain or are unreliable in service. Motors, on the other hand, are always obtainable locally and are very robust in operation. This book is intended to help local manufacturers and rural development engineers to select a motor and convert it for use as a generator for a micro-hydro scheme.

The development of this technology is the fruit of collaborative efforts between ITDG staff and colleagues and friends in Sri Lanka, Nepal and Indonesia, where the field work has been carried out. It has also been an excellent example of what can be achieved by matching the resources of UK research institutions to needs in developing countries.

The use of motors as generators is now well proven and promises to be an important element in establishing self-sustaining local capacity for village-scale hydro in developing countries.

Take-up of the technology outlined in this book is now freely available to individuals, communities and organizations overseas, helping to fulfil ITDG's strategic aim of wide dissemination of appropriate technologies, which can be manufactured locally at affordable cost. ITDG continues to widen the availability of appropriate technologies through its on-going programme of training courses in micro-hydro and other aspects of rural energy. Specific courses are also held on induction generators and local manufacture of electronic controllers. For details, please write to ITDG at the Rugby address.

ITDG Energy Unit
Intermediate Technology Development Group (ITDG)
Myson House, Railway Terrace, Rugby CV21 3HT, UK

ACKNOWLEDGEMENTS

I would like to express my thanks to the Intermediate Technology Development Group (ITDG) and the Overseas Development Administration (ODA) Engineering Division for funding publication of this book, as well as much of the research and development work that has made it possible. In particular, their open technology approach has made it possible for technical data to be documented and freely transferred throughout the course of this work, and has included several training courses for local engineers and manufacturers.

I would also like to acknowledge The Nottingham Trent University for providing research facilities and technical and financial support.

In addition, I am grateful to Ian Macwhinnie, Arthur Williams, Keith Pratt and Doreen Smith for reading the manuscript and making valuable suggestions.

Thanks are also due to Brook Crompton for permission to publish their motor data, the photograph of an induction motor (Figure 1) and the technical drawing (Figure 4).

Nigel Smith
Nottingham 1994

INTRODUCTION

Until recently, micro hydro installations have always used synchronous generators to produce a.c. electricity if no grid supply is available. Often, for reasons of availability and low cost, machines designed for use with petrol or diesel sets are used. These machines are generally designed for intermittent use, under direct drive conditions. It is little wonder that they do not last long in continuous operation, especially when belt driven by a water turbine with a runaway speed considerably higher than its normal operating speed. Quality synchronous machines are built for such arduous duties, though they are less easy to obtain and are expensive.

In recent years, induction motors have been installed as generators on micro hydro schemes. The induction motor is the most common type of electrically powered prime-mover used by industry. Induction generators are sometimes referred to as IGs or IMAGs - induction motors as generators. In Nepal alone more than one hundred induction generators have been installed in the past six years, and they have proven to be considerably more reliable than the available synchronous generators.

With early schemes, because of the absence of a suitable control system, a near constant load had to be maintained at all times. This was generally achieved by having a single switch or contactor next to the generator and permanently connecting all the loads to the supply. A low cost induction generator controller (IGC) has now been developed, and locally manufactured IGCs are fitted to all but the smallest of schemes. These allow a wide variety of loads to be connected and disconnected as required.

It has been assumed that the reader has a basic understanding of single and three-phase circuits and induction motors. Those without the necessary background knowledge should read the relevant chapters of standard diploma/certificate level electrical engineering books.

For consistency, a supply voltage of 240 V single-phase and 415 V three-phase and a frequency of 50 Hz has been assumed throughout the book. However, the principles taught can be applied to all other supply specifications.

Figure 1 Induction motor (Brook Crompton)

1. ADVANTAGES AND DISADVANTAGES OF INDUCTION GENERATORS

The main advantages and disadvantages of using induction machines instead of synchronous generators for stand-alone micro hydro are as follows.

Advantages

- **Availability**

Induction motors are much more widely available than synchronous generators. In some cases, second-hand machines can be reconditioned in order to reduce costs.

- **Cost**

Induction generators, including their excitation capacitors, are generally cheaper than synchronous generators. This is especially true for low power ratings. For example, a 10 kW induction generator is typically half the cost of a synchronous generator. This price difference varies between countries, since it depends on whether local manufacture takes place and on the sources of imported machines. In some countries there is little or no financial advantage to using induction generators for schemes of more than 25 kW, whereas in others there is an appreciable cost saving.

- **Robustness**

Induction machines are very robust and have a simple construction, as shown in Figure 1. They have no winding, diodes or slip rings on their rotor. Solid, normally cast bars, replace the rotor winding and enable the rotor to withstand considerable overspeed. In addition, the machines are normally totally enclosed, ensuring good protection against dirt and water. They are designed for continuous operation with belt drives under arduous industrial conditions.

Disadvantages

- **Voltage rating**

Induction machines are not always available with suitable voltage ratings for use as generators. Modification to the winding connections, or in extreme cases rewinding, may be required.

- **Calculation required**

Whilst synchronous generators can be purchased ready for use, the induction machine will not work without capacitors of a suitable value being fitted.

- **Motor starts**

Motors are more easily started with synchronous generators than induction generators. Induction motors, with a capacity that is large when compared to the generator rating, can cause severe voltage dips or even loss of excitation when started from induction generators.

This book explains how to choose and if necessary modify the winding arrangement of an induction machine in order to make it generate at the desired voltage, and how to calculate the capacitance required. It also advises on ways of improving motor starting and the limits of motor starting capability. It explains how to use a three-phase motor as a single-phase generator, and describes the protection and control requirements.

2. INDUCTION MACHINE CONSTRUCTION AND OPERATION

Construction

The mechanical features of an induction machine can be seen from Figure 2. From an electrical point of view, the induction machine consists of two parts: a fixed, wound stator core on the outside and a rotor that rotates in the centre. The stator winding consists of coils of insulated copper wire fixed into slots in the core to form a distributed winding of a similar type to that used with a synchronous generator.

The induction machine rotor is very different from that of a synchronous generator. The standard squirrel cage rotor core is cylindrical and built up from thin sheets of steel into which slots have been punched for the conductors. The conductors generally consist of aluminium bars that are short-circuited at each end by aluminium rings. For small machines, the bars and end-rings are cast in one operation using the rotor core as part of the die. For large machines, copper or aluminium bars are welded or brazed to the end-rings.

Operation

In order to understand the operation of a stand-alone, i.e. non grid connected, induction generator it is easiest to take motor operation as a starting point.

Induction motor operation

When an induction machine is connected to an a.c. supply, magnetizing current flows from the supply and creates a rotating magnetic field in the machine.

The rotating field cuts the short-circuited rotor bars, inducing currents in them which, because they are flowing in the magnetic field, react with it producing a torque. This torque drags the rotor round with the field, but at a slightly lower speed. The small difference in speed arises because without it no currents would be induced in the rotor and, therefore, no torque would be produced to turn it. When a load is applied to the motor the speed difference will increase as a greater torque must be produced.

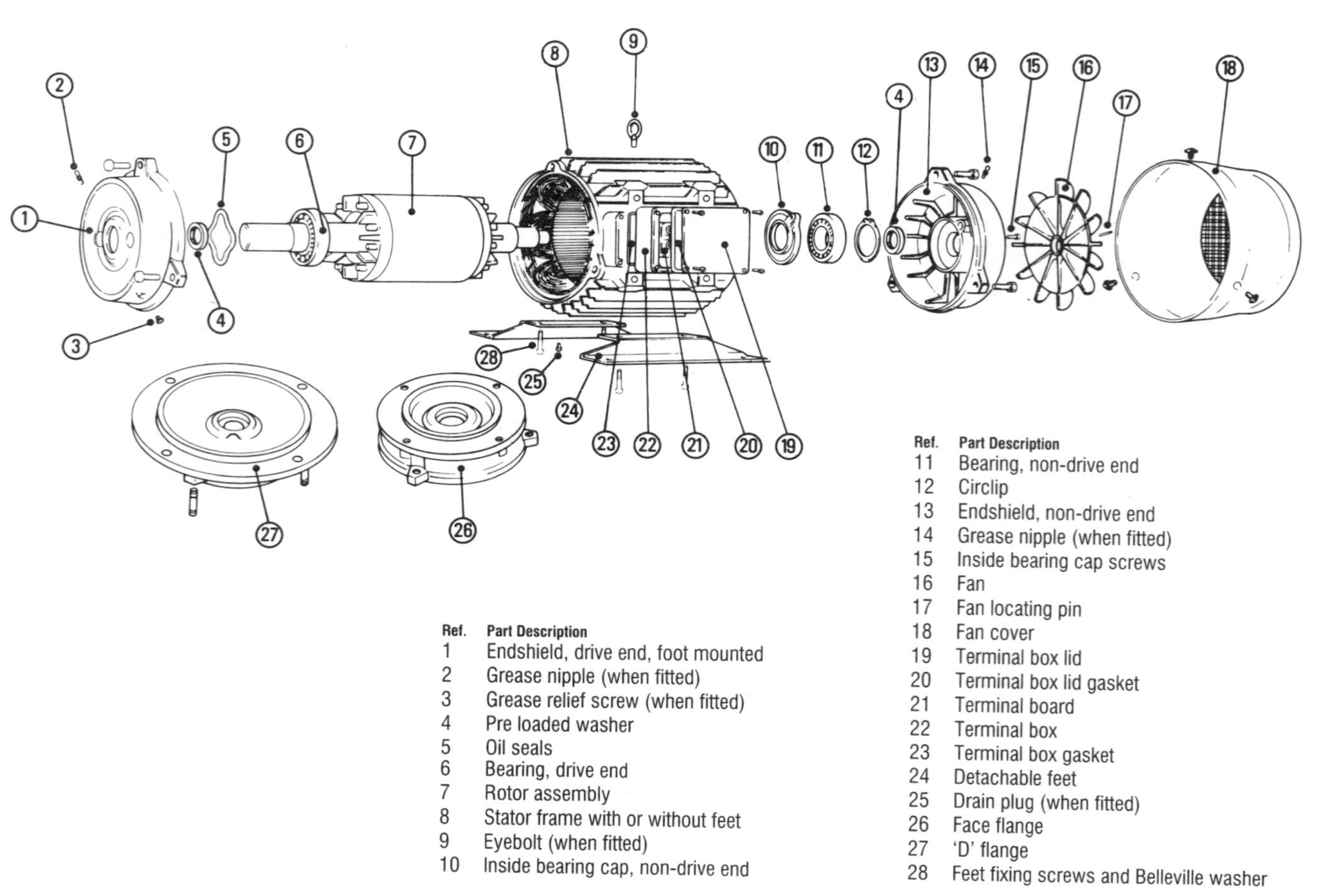

Ref.	Part Description
1	Endshield, drive end, foot mounted
2	Grease nipple (when fitted)
3	Grease relief screw (when fitted)
4	Pre loaded washer
5	Oil seals
6	Bearing, drive end
7	Rotor assembly
8	Stator frame with or without feet
9	Eyebolt (when fitted)
10	Inside bearing cap, non-drive end
11	Bearing, non-drive end
12	Circlip
13	Endshield, non-drive end
14	Grease nipple (when fitted)
15	Inside bearing cap screws
16	Fan
17	Fan locating pin
18	Fan cover
19	Terminal box lid
20	Terminal box lid gasket
21	Terminal board
22	Terminal box
23	Terminal box gasket
24	Detachable feet
25	Drain plug (when fitted)
26	Face flange
27	'D' flange
28	Feet fixing screws and Belleville washer

Figure 2 Exploded view of an induction generator

The difference between the speed of the rotor and the speed of the rotating field is called the 'slip' and is defined as:

$$\text{Slip,} \quad s = \frac{(n_s - n_r)}{n_s} \qquad (1)$$

Where n_s is synchronous speed (the speed of the rotating field)
n_r is the rotor speed

Without load connected, the slip of the induction motor will be very small, less than 0.01 (that is 1%). For a machine of 1 kW the full load slip will be about 0.05 (or 5%). Larger machines have smaller slips. For more information see the speed column in Table 1.

Supply-connected induction generator operation

If the same supply-connected induction machine is now driven at above synchronous speed, so that the slip becomes negative ($n_r > n_s$), a torque is supplied to the rotor rather than by the rotor and the machine acts as a generator, supplying power to the network. However, it still takes its magnetizing current from the supply in order to create the rotating field, just as though it were a motor. The full load power output is achieved at a slip of similar value (but negative) to the full load motoring slip. Example 1 illustrates the difference in slip and shaft speed between motor and generator operation.

Stand-alone induction generator operation

The magnetizing current of an induction machine can be supplied, in total or in part, by capacitors. In fact, capacitors are often fitted to large induction motors and supply-connected induction generators to reduce the reactive current drawn from the supply, especially when the electricity company imposes charges for poor power factor.

In the case of a stand-alone induction generator, the capacitors are the only external source of magnetizing current. Therefore, in order to obtain the required operating voltage at the desired frequency, the amount of capacitance must be carefully chosen.

kW	Poles	Full load speed (rpm)	FLC* (A) at 415 V	Efficiency			Power Factor		
				FL	3/4L	1/2L	FL	3/4L	1/2L
0.55	2	2810	1.32	74	76	76	.79	.68	.51
	4	1400	1.50	73	72	69	.70	.61	.47
	6	900	1.65	73	72	69	.64	.54	.42
0.75	2	2850	1.75	78	78	74	.77	.73	.61
	4	1400	1.95	77	77	74	.70	.60	.48
	6	920	2.40	71	70	65	.62	.53	.41
1.1	2	2850	2.5	79	78	73	.80	.74	.64
	4	1410	3.0	79	78	71	.65	.55	.43
	6	900	3.1	75	75	74	.66	.58	.47
1.5	2	2850	3.2	76	78	70	.86	.82	.73
	4	1420	3.6	82	83	80	.71	.61	.47
	6	940	4.3	75	74	72	.65	.55	.44
2.2	2	2850	4.4	81	81	80	.86	.82	.73
	4	1420	4.9	80	81	80	.78	.71	.58
	6	945	5.9	77	77	75	.68	.58	.47
3.0	2	2860	5.8	83	83	81	.88	.83	.70
	4	1420	6.5	81	82	80	.80	.74	.61
	6	950	6.4	83	84	82	.78	.73	.61
4.0	2	2840	7.2	85	86	85	.90	.88	.81
	4	1420	8.4	83	84	83	.80	.72	.61
	6	960	9.2	84	83	82	.72	.62	.50
5.5	2	2900	10.6	85	85	83	.85	.83	.76
	4	1445	11.3	85	86	85	.80	.74	.61
	6	955	12.8	84	85	83	.71	.63	.50
7.5	2	2900	14.1	86	86	85	.87	.84	.77
	4	1450	14.6	86	86	85	.83	.76	.64
	6	970	15.6	87	87	85	.77	.70	.55
11	2	2930	20	87	87	85	.87	.83	.74
	4	1460	20	88	88	87	.86	.81	.70
	6	970	24	88	88	87	.73	.65	.55
15	2	2940	27	88	88	87	.89	.88	.85
	4	1470	27	90	90	89	.87	.80	.68
	6	970	29	88	88	87	.83	.80	.70
18.5	2	2950	32	90	90	87	.89	.87	.75
	4	1460	33	89	89	88	.89	.88	.83
	6	970	36	90	90	88	.81	.76	.67
22	2	2920	38	88	88	86	.92	.91	.88
	4	1460	38	90	91	90	.89	.86	.81
	6	970	41	91	91	89	.83	.78	.70
30	2	2945	53	89	88	86	.89	.87	.79
	4	1470	54	91	91	89	.86	.83	.76
	6	970	55	91	91	90	.83	.78	.70
37	2	2950	64	90	89	86	.89	.85	.80
	4	1465	66	91	91	89	.86	.80	.72
	6	975	68	92	91	89	.83	.78	.70
45	2	2950	77	90	89	87	.91	.90	.84
	4	1465	79	91	91	89	.87	.81	.74
	6	975	81	92	92	90	.84	.79	.71
55	2	2955	93	90	89	87	.91	.90	.84
	4	1470	98	91	91	89	.83	.78	.68
	6	985	96	92	92	90	.86	.81	.74

Table 1 Motor performance data (Brook Crompton)

* Where FLC is the full load current.

Example 1

For the 4-pole 7.5 kW induction machine listed in Table 1:

a) What is the full load slip?
b) At what shaft speed will the machine operate as a generator when running at full load and 50 Hz?

a) The full load motor shaft speed is given as 1450 rpm.

The synchronous speed is calculated from:

$$n_s = \frac{120 \times f}{p} \qquad (2)$$

Where f = supply frequency
p = number of poles

In this case, $$n_s = \frac{120 \times 50}{4} = 1500 \text{ rpm}$$

Hence, using Equation 1, $$\text{slip, } s = \frac{1500 - 1450}{1500} = 0.033$$

b) Since full load generating slip is approximately equal in magnitude to full load motoring slip, but negative, $s = -0.033$

Re-arranging Equation 1, $$n_r = n_s(1 - s)$$

$$n_r = 1500\left(1 - [-0.033]\right) = 1550\text{rpm}$$

Hence, instead of the shaft speed being 50 rpm below the synchronous speed, as a generator the shaft speed is 50 rpm above the synchronous speed.

For a build-up of voltage to occur, sufficient remanent magnetism must be present in the rotor. Remanent magnetism is the initial magnetism present in the rotor steel. It is generally sufficient to produce a small voltage, of about a volt, at synchronous speed with no capacitance connected. There may be insufficient magnetism if, since it was last used, the machine received a large impact or generation was collapsed with a resistive load connected. Remanent magnetism is dependent upon the type of steel used. Low loss steels, as used in energy efficient induction machines, tend to have the lowest levels of remanent magnetism.

If there is insufficient magnetism to excite the generator at rated frequency, increase the generator speed, because at higher frequency less magnetism is required for excitation to occur. In almost all cases, this will be sufficient to excite the generator. However, if this fails, remanent magnetism can be increased by connecting a d.c. supply for a few seconds across any two of the machine terminals, before running the machine up to speed. A car battery is more than sufficient for this purpose. Ordinary dry cells, connected in series, can also be used. High capacity dry cells, such as those used in large torches, are best as they can source the high current required.

A simplistic but useful way to understand the basic operation of a stand-alone induction generator is to represent the machine simply by its magnetizing reactance. The highly simplified equivalent circuit shown in Figure 3 can then be used. This is a fairly accurate representation for the purpose of determining capacitor requirements.

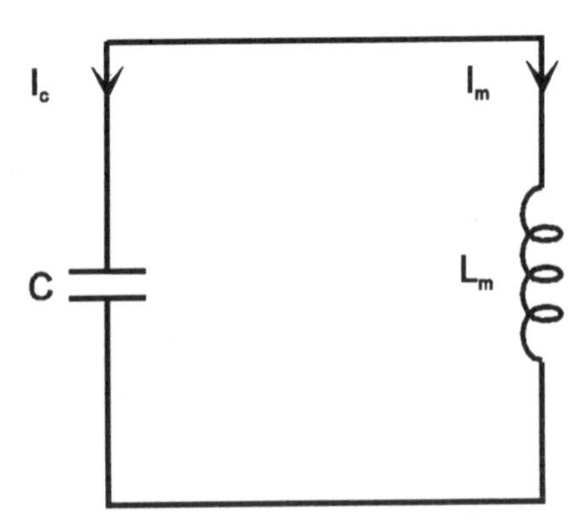

Figure 3 Highly simplified induction generator equivalent circuit

With the shaft rotating, current will begin to flow due to the remanent magnetism present in the rotor. The capacitor current, I_c, will equal the magnetizing current, I_m, and the machine and capacitors will act as a resonant circuit at an angular frequency, ω, fixed by the shaft speed of the machine. Provided that sufficient capacitance is present, the current will rapidly increase until stable operation is reached when the impedance of the capacitors equals the magnetizing impedance as given by Equation 3.

$$\frac{1}{\omega C} = \omega L_m \qquad (3)$$

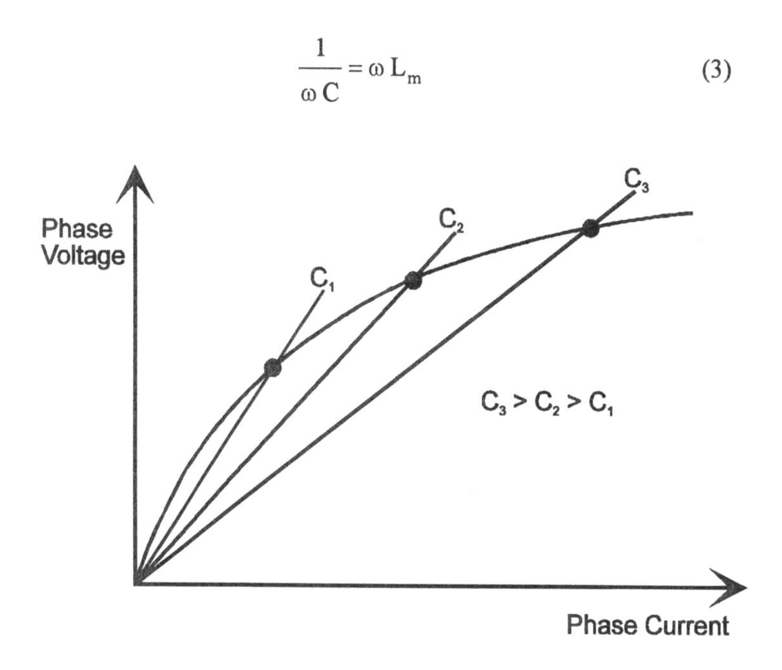

Figure 4 *No load excitation characteristic for an induction generator driven at constant speed (for three values of excitation capacitance)*

Stable operation occurs because the magnetizing inductance is a non-linear function of current, due to magnetic saturation of the rotor and stator steel. Provided that sufficient capacitance is connected, the voltage against current characteristic for the capacitor(s) meets the voltage against current characteristic for the magnetizing inductance, as shown in Figure 4. This is the operating point of the generator. Increasing the capacitance will increase the operating voltage but, since more current flows into the machine, additional power will be lost as heat in the stator windings.

3. SELECTION OF AN INDUCTION MOTOR FOR USE AS A GENERATOR

The number of options available when selecting an induction machine will depend upon the accessibility of manufacturers and suppliers and the range of machines that they stock. The following guidelines cover the broadest range of options that may be open to the purchaser, though some of the options are rarely available.

Rotor type

Induction motors with cage rotors are by far the most common type of machine and should always be selected. Some manufacturers also produce wound rotor machines, but these are more expensive and less robust and should, therefore, be avoided. Wound rotor machines can be recognized by the fact that they have additional terminals which connect to the rotor by means of slip rings.

Site conditions

Motors should have an IP classification on their nameplate which indicates their level of protection against penetration by solids and liquids. The first number refers to penetration by solids and the second to penetration by liquids. The levels of protection, or IP numbers, most commonly available are:

IP 21 Protected against solids greater than 15 mm and against vertically falling dripping water.

IP 22 Protected against solids greater than 15 mm and against dripping water falling at any angle up to 15° from the vertical.

IP 23 Protected against solids greater than 15 mm and against water falling as a spray at any angle up to 60° from the vertical.

IP 44 Protected against solids greater than 1 mm and water splashing from any direction.

IP 54 Protected against dust and against water splashing from any direction.

IP 55 Protected against dust and against water jets from any direction.

Most motors are classified as Totally Enclosed Fan Ventilated (TEFV) and are protected to IP54 or IP55. These machines are suitable for use in dusty environments, such as mills. For direct-drive applications IP55 motors are best because of the additional protection that they provide.

Induction motors with a capacity above 10 kW are sometimes available with drip-proof frames (IP21, IP22 or IP23). They are cheaper than TEFV machines and are, therefore, worth considering, except in dusty environments.

Lubrication

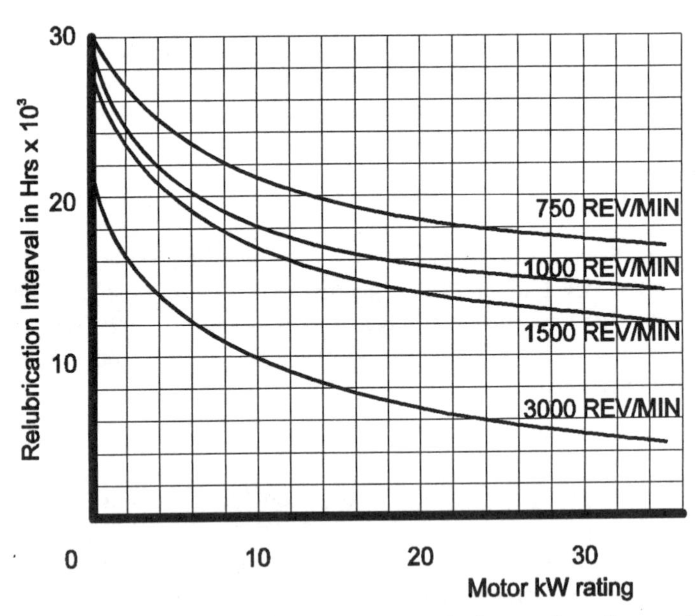

Figure 5 Typical relubrication intervals for good quality machines

As shown in Figure 5, induction machine bearings do not require frequent lubrication. The relubrication interval depends upon the speed and power rating of the machine. Because relubrication intervals are long, manufacturers often do not fit regreasing facilities to their motors. This is acceptable if the machine is run for only a few hours per week but, for generators operating continuously, the machine will need to be dismantled to regrease the bearings at regular intervals. It is best to avoid dismantling the machine, especially in dusty environments, and much easier to apply grease using a grease gun. For these reasons, motors fitted with regreasing facilities should be used whenever possible. Grease relief should also be specified to prevent overgreasing.

Manufacturer's instructions for regreasing should be followed if these are available. Otherwise relubrication should be carried out at the intervals shown in Figure 5 using the following procedure:

1) Wipe clean the grease gun fitting and regions around the motor grease fittings.

2) Remove the relief plugs.

3) To each bearing add a small quantity of grease compatible with the bearing size.

4) Allow the motor to run for about ten minutes before refitting the relief plugs so that excess grease can be expelled.

Frame type

Manufacturers often provide the option of either foot or flange mounting, as shown in Figure 2. For certain applications such as the direct-drive pelton, shown in Figure 6, flange mounting may be preferable.

Insulation and temperature rise

The lifetime of the machine windings is dependent upon the number of hours per day that the machine is used, the quality of the insulation and the operating temperature. A rough guide to the effect of temperature upon insulation life is that for every 10°C reduction in operating temperature the insulation life of the windings will be doubled. The most common classifications of insulation in current use are:

Class B permitting a winding temperature rise of 80°C above a 40°C ambient, or 120°C maximum.

Class F permitting a winding temperature rise of 100°C above a 40°C ambient, or 140°C maximum.

Class H permitting a winding temperature rise of 110°C above a 40°C ambient, or 150°C maximum.

Figure 6 *Directly coupled pelton turbine-induction generator, Mango, N. Pakistan*

If the windings are operated at the maximum temperature for their classification, then, according to insulation standards, the life of the insulation will be a minimum of 20,000 hours. In practice the life is much longer, since the machine is unlikely to operate at its maximum temperature rise and in maximum ambient temperature all the time.

Improvements in insulation over the years have meant that machines can run hotter, allowing thinner conductors to be used. As a result, machines have become smaller and cheaper. However, using thinner conductors reduces efficiency and nowadays it is the trade-off between efficiency and machine cost, rather than maximum operating temperature, that tends to determine conductor size. The cost of class F insulated wire is now little more than for class B. Hence, many modern induction machines are wound with class F wire even if the operating temperature does not necessitate its use. Indeed, some manufacturers use class F insulation but state that their motors are designed to operate within class B limits. Such motors will have approximately four times the winding life of the same machines fitted with class B insulation.

Since most induction generators are used continuously and down time for rewinding is very inconvenient, insulation life is an important consideration and motors with high temperature insulation should be sought.

In rare cases the generator output will need to be derated because of operation at high altitude. Standard ratings apply to temperatures not exceeding 40°C at altitudes up to 1000 metres above sea level. The following correction factors should be used at higher altitudes.

Altitude	2000m	3000m	4000m
When ambient temperature is 40°C reduce rated output by:	92%	85%	75%
Maximum ambient temperature for full rated output	32°C	24°C	16°C

Table 2 Output derating for high altitude operation

Efficiency

Induction machines have a slightly lower efficiency when operated as generators than as motors, as shown in Appendix 2. Often the only option for improving efficiency is to change the operating voltage and frequency, as discussed in Chapter 5. However, some manufacturers produce a range of high efficiency motors by increasing the amount of copper used in the windings and the quantity and/or quality of the steel used in the stator and rotor cores. Where available, these machines should be considered when selecting a suitable generator. They are typically 50% more expensive than a standard motor, but this cost is often justified by the increased power output, as shown in Example 2. Energy efficient machines have better efficiencies than standard machines at part-load as well as full load.

The low-loss steels used in energy efficient machines result in lower remanent magnetism. To provide sufficient magnetism for excitation to occur a battery may well be required, as explained in the previous chapter.

Power rating

The greater the power drawn from the generator, the higher its operating temperature and hence the shorter the life of its windings. To ensure long winding life when used as a generator, the machine should be kept below its full load operating temperature as a motor.

Winding temperatures for an induction machine operating as a motor and as a generator are given in Appendix 2. By careful choice of the operating conditions in generator mode, namely voltage and frequency, a generator mode power rating similar to the motor mode power rating can be achieved with an equivalent temperature rise. However, this assumes balanced operation and near optimum operating conditions. Since maintaining an equal load on each phase is rarely possible on small generating systems and operation under optimum conditions is not always possible, induction machines should be derated when used as generators. A derating factor of 0.8 is recommended and is also applicable to single-phase generation from a three-phase machine, as explained in Chapter 7.

A de-rating factor of 0.8 provides a good safety margin, since heat transfer between the stator windings helps to correct for temperature differences due to unequal currents in the windings.

Example 2

A micro hydro engineer visits a motor supplier in order to purchase a 7.5 kW motor for use as a 6 kW generator. The supplier offers him a standard machine, with 86% efficiency, for $400 or an energy efficient machine, with 90% efficiency, for $600.

a) If the standard machine will give an electrical output of 6 kW when driven by the turbine at the micro hydro site, what will be the output if the energy efficient machine is used?
b) What is the cost per kW of the additional power produced?

a) The turbine power output is:

$$P_{turb} = \frac{P_{gen}}{\eta_{gen}} \qquad (4)$$

Where P_{gen} is the generator power output

η_{gen} is the generator efficiency

Entering the standard induction machine efficiency,

$$P_{turb} = \frac{6}{0.86} = \underline{\underline{6.98\,kW}}$$

For the energy efficient generator,

$$P_{gen} = P_{turb} \times \eta_{gen} = 6.98 \times 0.9 = \underline{\underline{6.28\,kW}}$$

b) An additional 280 Watts has been produced at an extra cost of $200. Hence, the additional cost per kilowatt is:

$$\frac{\$200}{0.28\,kW} = \underline{\underline{720\$/kW}}$$

The decision as to whether to use an energy efficient induction machine will depend upon the cost per kW of the extra power produced, compared to the cost per kW of the complete hydro scheme, and the usefulness of the additional power produced.

Avoid using a generator larger than required, as induction machines, especially of smaller capacities, have poor part load efficiencies, as shown in Appendix 2.

Voltage rating

When selecting an induction machine the voltage rating is a very important consideration. In some cases it will be possible to obtain a machine that can be used directly as a generator. In other cases some modification to the windings will be necessary.

Small generators
Induction machine manufacturers often wind machines of up to 3 kW capacity to provide the choice of either 240 V delta or 415 V star operation. These machines will be labelled 240/415 V on the nameplate. The star and delta winding configurations are shown below. The 240 V connection will be used when operating the machine as a single-phase generator, as explained in Chapter 7. The 415 V connection is suitable for three-phase generation, as the presence of a star point, which can be earthed and forms the neutral connection, enables single-phase (240 V) loads to be connected between any line and neutral, and three-phase loads to be connected to the three lines.

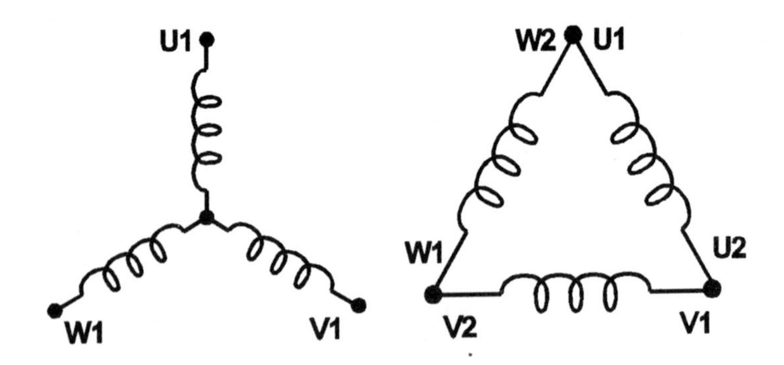

Figure 7 Star and delta winding configurations

Motors that can be connected for either 240 V or 415 V operation are supplied with all six ends of their windings available in the terminal box. This enables easy connection in star or delta as shown in Figure 8. Some small motors are supplied with the windings in star and with the star connection made internally so that just three connections are provided at the terminal box. In such cases, it is a simple job for a motor repairer to provide a connection to the star point at the terminal box. If a spare terminal is not available a good earthing point to the machine case, within the terminal box, can be used for the star point and shared with the earth connection. Reconfiguration in delta again requires internal reconnection of the windings and the delta connected lines must be brought to three terminals in the terminal box. Motors wound for 415 V delta connection should be avoided, as explained in the next section.

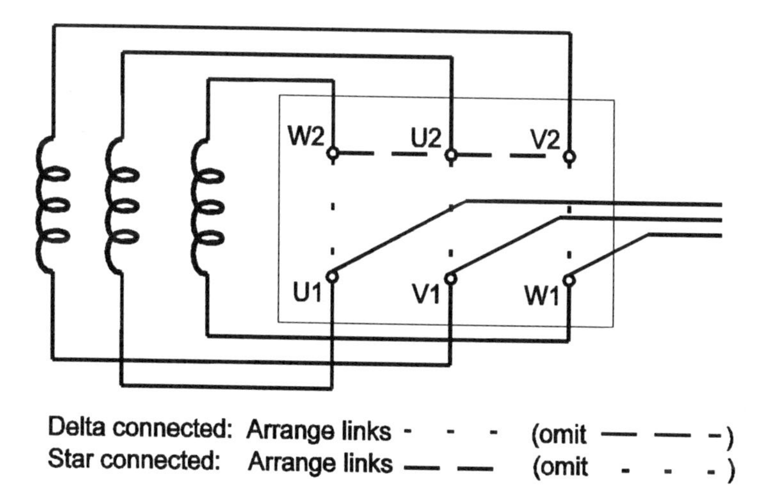

Figure 8 Star and delta connections at the terminal block

Larger generators

Induction motors with a capacity above 3 kW are generally wound for star-delta starting. The machines are designed to run with a 415 V delta connection as explained in Chapter 10.

From a generating point of view, a 415 V delta connection has the disadvantage that there is no star point to serve as a neutral and earth connection. Hence, single-phase (240 V) loads cannot be connected and there are significant disadvantages in terms of earthing. This winding arrangement can be used to advantage in situations where there is a long transmission line, as explained in Appendix 3. However, for most schemes 415 V delta is unsuitable and a star connected machine should be used.

There are three possibilities for obtaining a 415 V star connected machine. These, in order of ease and cost, are as follows:

1) Deal directly with the manufacturer or his agent in order to purchase a machine wound for 240/415 V operation. Most manufacturers will supply such machines at little or no extra cost, though the delivery time is usually increased.

2) Reduce the operating voltage by reconnecting the groups of winding coils to a parallel configuration. This can be achieved because machines are generally wound with an even number of coil groups per phase, connected in series. Reconnection in parallel to yield 207.5 V, as explained in Appendix 4, is a relatively straightforward operation, since only the ends of the coil sets need reconnecting and no rewinding is required. To operate the machine at 240 V, an increase in operating frequency of approximately 10% will be required to compensate for the increased saturation, as explained in Chapter 5.

3) Have a 415 V delta connected machine rewound for 415 V star. This is expensive as it costs approximately half the price of a new machine. It is a good option if reconditioning a second-hand machine that must be rewound anyway. Make sure that you use a rewinder who can scale the turns and cross-sections of the cables correctly as many are used to just copying the existing winding.

If the first or third options are taken then it is worth considering winding the machine for a higher voltage in order to improve efficiency, as explained in Chapter 5.

Speed rating / drive arrangement

Ideally the generator should be directly driven by the turbine. This has advantages in terms of increased efficiency, reduced drive system costs, lower maintenance and simpler installation. Unfortunately, turbine speeds are often much slower than standard generator speeds and therefore this approach is not always possible. It may be suitable for high head or low flow installations or where high specific speed turbines such as pumps as turbines are appropriate.

Figure 9   Crossflow turbine driving induction generator, Tamgas, Nepal

Figure 10   1.5 kW geared induction motor (Paul Bromley)

Induction machines have the advantage over synchronous generators that 6 pole and sometimes 8 pole machines are available. However, for a given power rating, the cost relative to that of a 4 pole machine is typically 1.5 times for a 6 pole machine and 2 times for an 8 pole machine. There is generally little price difference between 2 pole and 4 pole machines.

With most crossflow installations a 4 pole machine is belt driven by the turbine, as shown in Figure 9. For very slow speed turbines, where a single stage belt drive is insufficient, geared induction machines, where available, are worth considering. Heavy duty gear boxes should be selected in order to ensure good reliability. Figure 10 shows a 20:1 geared 4 pole induction motor which is used with a water wheel in England. A tractor differential, already on site, is used to provide the first stage of the drive, though it would have been possible to purchase a geared motor for the full 120:1 step up required.

Although standard induction machines are designed for belt drive, care must be taken to limit the radial load in order to ensure good bearing life. Where possible, the motor supplier should be contacted for information on pulley dimension limits and the belt manufacturer's catalogue consulted for belt tension requirements.

4. EXCITATION CAPACITOR REQUIREMENTS

Voltage and current relationships in three-phase circuits

A good understanding of the voltage and current relationships in star and delta connected circuits is required in order to calculate excitation capacitance values correctly. The essential terms and relationships are given in Appendix 1.

Capacitor connection

For a three-phase generating system, the capacitors can be either star or delta connected, as shown in Figure 11. In the case of the star connected system, the star point of the capacitors should not be connected to the generator and system neutral as waveform distortion and increased losses will occur.

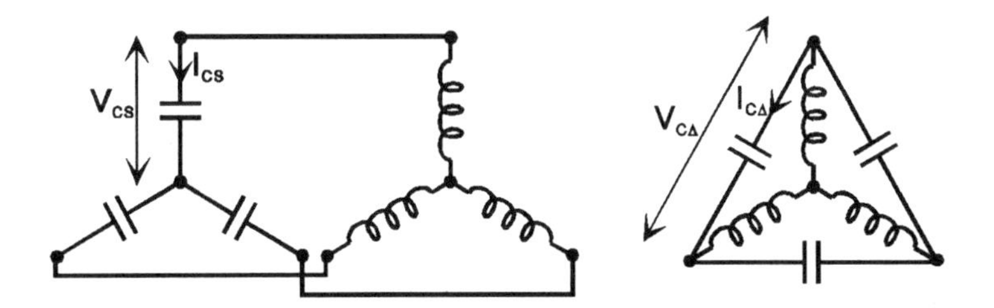

Figure 11 Star and delta connection of excitation capacitance

The star connection and delta connection are related as follows:

$$V_{C_\Delta} = \sqrt{3}.V_{C_S} \quad \text{and} \quad I_{C_\Delta} = I_{C_S} / \sqrt{3}$$

$$X_{C_\Delta} = \frac{V_{C_\Delta}}{I_{C_\Delta}} = \frac{\sqrt{3}.V_{C_S}}{I_{C_S}/\sqrt{3}} = \frac{3.V_{C_S}}{I_{C_S}} = 3.X_{C_S}$$

$$\text{Since } C = \frac{1}{\omega . X_C}, \quad C_\Delta = \frac{C_S}{3} \tag{5}$$

If the capacitors are connected in star, then three times as much capacitance is required than for delta connection, though lower voltage and therefore cheaper capacitors may be used. A sample calculation is presented in Example A2 of Appendix 1.

Selection of capacitors

Generally the most suitable type of capacitor to use is the 'motor run' type, used with some single-phase induction motors. They are widely available in sizes up to and sometimes above 40 micro-Farads (μF). Their voltage rating is usually 380-415 V, though sometimes 220-240 V types are also available. Capacitor prices are approximately proportional to their voltage rating. Hence, higher voltage capacitors are generally better value, since the volt-amp capacity increases as the square of the voltage.

'Motor run' type capacitors are generally quite cheap and, even for a star connection, should be less than 30% of the generator cost. 'Motor start' capacitors must be avoided, since they are not designed for continuous use. They are normally labelled 'motor start' or electrolytic and can also be recognized by their very low cost and large capacitance rating for their size. Capacitors designed for power factor correcting fluorescent lamps can be used with induction generators. However, they are rarely available in sizes above 10 μF.

Capacitor life is governed by quality of manufacture and operating voltage, frequency and temperature. The lifetime is very dependent upon voltage. For example, a capacitor that lasts a year at its rated voltage may last for thirty years at half rated voltage. Therefore, it is recommended that 415 V capacitors are used even for star connected arrangements, and that 415 V operation is avoided unless the capacitors are rated for more than 415 V.

Capacitor life is also dependent upon frequency. An increase in frequency will increase capacitor current, adding to the losses and thereby raising the temperature of the insulation.

Capacitor life is also highly dependent upon ambient temperature. The capacitors should be placed in a box that has ventilation holes or slots to help cool the capacitors, and this should be placed in a cool place away from direct sunlight.

Capacitors are only available in standard sizes and are generally specified with a tolerance of +/- 10%. Hence, without measuring individual capacitors it is difficult to obtain the precise capacitance required. The next section shows that great accuracy is not required, though efforts should be made to be within 10% of the value calculated.

Calculation of excitation capacitance

A precise calculation of the capacitance required to generate a given voltage under specific load conditions is only possible with an accurate knowledge of the electrical parameters of the induction machine, including the variation of the parameters with voltage. These parameters can be obtained by means of a number of standard tests, but expensive equipment is required. In practice it is sufficient to calculate an approximate value of excitation capacitance and adjust the turbine speed until the required system voltage is obtained. This will mean that the operating frequency may differ from the rated frequency of the induction machine, which is acceptable provided that the frequency is kept within reasonable limits. These limits are explained in Chapter 5.

Two methods are recommended for the approximate calculation of the excitation capacitance required: an electrical test method and a technique requiring only induction motor performance data.

No load test

The electrical test method involves a 'no load test' either as a motor or a generator. No load test results can be used to calculate excitation capacitance because the apparent power drawn by the induction machine at no load is approximately equal to the reactive power required when running as a generator at close to full load.

The motoring test is simplest, but it requires a three-phase supply with sufficient capacity to start the machine (the starting current will be approximately six times the rated full load current). A voltmeter and clip-on ammeter are required to measure the line voltage and current, in order to determine the apparent power and therefore the reactive power to be supplied by the excitation capacitors. A sample calculation is given in Example 3.

Example 3

A 2.2 kW, 4 pole, 50 Hz, 415 V, 3 phase induction motor draws a current of 3.5 A when supplied at its rated voltage and frequency and run as a motor with no mechanical load. Calculate the excitation capacitance which must be connected in star to make the machine generate at approximately its rated voltage when driven at slightly above its rated speed.

The total apparent power at no load,

$$\Sigma S_{no\,load} = \sqrt{3} \times V_{line} \times I_{line}{}^{*} = \sqrt{3} \times 415 \times 3.5 = 2516\,\text{VA}$$

Using the relationship that the no load apparent power equals the reactive power to be provided by the excitation capacitors (as given in the no load test section), the total reactive power,

$$\Sigma Q = \Sigma S_{no\,load} = 2516\,\text{VAR}$$

Hence, the reactive power per phase,

$$Q_{phase} = \frac{Q}{3} = \frac{2516}{3} = 837\,\text{VAR}$$

For star connected capacitors,

$$V_{phase} = \frac{V_{line}}{\sqrt{3}} = \frac{415}{\sqrt{3}} = 240\,\text{V}$$

$$I_{phase} = \frac{Q_{phase}}{V_{phase}} = \frac{837}{240} = 3.5\,\text{A}^{**}$$

Since, $$X_c = \frac{V}{I} = \frac{1}{2\pi fC}, \quad C = \frac{I}{2\pi fV} \tag{6}$$

Hence, $$C = \frac{3.5}{2\pi.50.240} = \underline{\underline{46\,\mu\text{F}}}$$

* For derivation see Appendix 1.

** Note that for a star connection the capacitor current is the same as the no load line current.

If there is no three-phase supply of sufficient capacity to start the induction machine, then the no load test can be performed with the machine operating as a generator. A small single or three-phase motor is used to belt drive the test machine. The test machine should be driven at very close to synchronous speed and capacitors connected to excite the machine to rated voltage. Provided the test machine is driven at very close to synchronous speed, this is more accurate than the motoring test though it takes much longer to set up. It is generally only worthwhile for old machines with no performance data.

The no load test method underestimates the capacitance required. However, it is quite accurate for machines below 5 kW.

Manufacturers' data

The second method uses manufacturers' data to estimate the reactive power required from the capacitors, by means of the full load current and power factor.

The capacitive reactance for full power factor correction can be obtained from this information and the rated voltage. If the manufacturers' data is unavailable then an estimate can be obtained using the data given in Table 1. This will be fairly accurate for machines of modern design. A sample calculation is given in Example 4. Further practice can be obtained by repeating this calculation for the machine of Example 3 and comparing the results.

Usually, if either of these two methods for capacitance calculation are used, the generator will produce too low a voltage when operated at rated frequency. The reason for this is that they do not fully compensate for the increased saturation when the machine is operated as a generator under load.

However, the use of extra capacitance to allow for the increased saturation is not recommended. There are two reasons for this:

1) A high level of saturation will reduce generator efficiency and maximum output, as shown in Example 5. There are two much more preferable alternatives: operation at reduced voltage and operation at increased frequency, as explained in Chapter 5.

2) Manufacturers' data is not always reliable. If too much capacitance is connected then under frequency operation will occur, which is inefficient for the generator and may damage loads. In general, too little capacitance is better than too much.

Example 4

A 15 kW, 4 pole, 50 Hz, 415/240 V induction motor is star connected for operation as a generator. Using the performance data given in Table 1, calculate the amount of excitation capacitance that should be connected in delta to make the machine generate at approximately its rated voltage, when driven at slightly above its rated speed.

From Table 1, the full load line current is 27 A and the power factor is 0.87. The total apparent power at full load,

$$\Sigma S = \sqrt{3} \times V_{line} \times I_{line} = \sqrt{3} \times 415 \times 27 = 19408\,\text{VA}$$

Hence, the real power,

$$\Sigma P = \Sigma S \cos\phi = 19408 \times 0.87 = 16885\,\text{W}$$

The reactive power can be obtained from the power triangle:

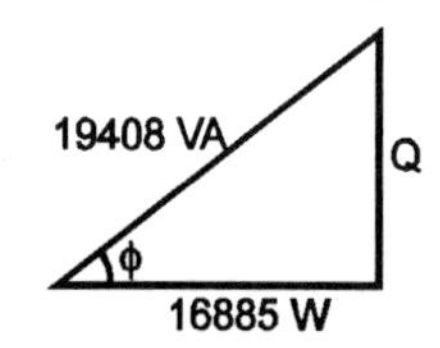

$$\Sigma Q = \sqrt{19408^2 - 16885^2} = 9569\,\text{VAR}$$

Hence, the reactive power per phase,

$$Q_{phase} = \frac{\Sigma Q}{3} = \frac{9569}{3} = 3189\,\text{VAR}$$

For delta connected capacitors,

$$V_{phase} = V_{line}$$

$$I_{phase} = \frac{Q_{phase}}{V_{phase}} = \frac{3190}{415} = 7.69\,\text{A}$$

From Equation 6, $C = \dfrac{I}{2\pi f V} = \dfrac{7.69}{2\pi.50.415} = \underline{\underline{59\,\mu\text{F}}}$

Example 5

When the 2.2 kW machine of Example 3 is installed on site it is found that to produce rated voltage at rated frequency the capacitance required must be increased by 30% compared to the initial calculation. Determine:

(a) The reactive power supplied by the capacitors.
(b) The maximum power output of the generator if the full load line current of the motor, given as 4.9 Amps, is not to be exceeded.

(a) The initial estimate for the total reactive power, $\sum Q$, was 2516 VAR. A 30% increase gives 3271 VAR.

(b) The maximum apparent power for the induction machine is set by the rated line current. Hence,

$$\sum S = \sqrt{3} \times V_{line} \times I_{line} = \sqrt{3} \times 415 \times 4.9 = 3522\,\text{VA}$$

The maximum real power can be obtained from the power triangle:

The maximum power output is,

$$\sum P = \sqrt{\sum S^2 - \sum Q^2} = \sqrt{3522^2 - 3271^2} = \underline{\underline{1306\,\text{W}}}$$

This is only 60% of the motor power rating and, since rated current is flowing for this reduced output, the efficiency will be significantly below the motoring efficiency. In practice, the power rating may have to be reduced further, since the calculation assumes equal loading on each phase of the generator.

It is a good idea to select combinations of capacitors that will allow some fine-tuning to be done on site. This can be used to ensure that the operating frequency is acceptable and to maximise the power output, as explained in Chapter 11.

5. OPERATING VOLTAGE AND FREQUENCY

The operating voltage and frequency of the generator will depend upon the turbine power output, the excitation capacitance and the load connected, including its power factor. There is some flexibility as to the choice of operating voltage and frequency which can be used to advantage to improve the efficiency of the system.

Operating voltage

Generator considerations

As explained in Chapter 2, an induction generator can be operated over a range of voltages. Figure 4 showed how, for constant speed operation, the voltage is determined by the amount of capacitance connected.

For operation at synchronous speed, the highest voltage that can be achieved is typically 125% of the motor rating and the lowest voltage approximately 65% of the motor rating. The upper limit is set by the voltage at which the excitation current equals the rated current of the machine and the lower limit is the voltage at which the machine is sufficiently saturated to operate stably. Operation at close to the upper voltage limit is not recommended, because, due to winding temperature considerations, the large current flowing into the generator from the excitation capacitors reduces the maximum load that can be connected and reduces the efficiency of the generator. This was shown in Example 5. Operation at close to the lower voltage limit is not recommended for two reasons:

1) The reduced excitation capacitance makes the generator very sensitive to reactive loads, which could cause significant variations in frequency and possibly loss of excitation, as explained in Chapter 6.

2) The efficiency will be below the optimum efficiency, particularly when the generator is heavily loaded. This is because, in order to achieve the same power output at reduced voltage, the load current must be increased. For example, a 33% reduction in voltage will require a 50% increase in load current in order to maintain power output. This increases stator and rotor heating and reduces efficiency.

As shown in Appendix 2, to achieve optimum efficiency the induction machine should be operated as a generator at 90-95% of the motor rating. In addition to improving efficiency, this cuts capacitor costs and improves insulation life.

When selecting the operating voltage for the generator, transmission line voltage drop must also be considered.

Load considerations

Voltages above the manufacturers' ratings should be avoided as the lifetime of the equipment, be it a heater, lamp or motor will be reduced. Manufacturers' data indicate that for a standard incandescent lamp a continuous overvoltage of just 5% will decrease the expected life of the lamp by approximately 50%. Hence, it is clear that operation of equipment at above rated voltage should be avoided.

The effect of an undervoltage with a heater or incandescent lamp is to increase the life and decrease the output. In the case of fluorescent lamps too low a voltage can prevent turn on and may cause the starter to fail due to repetitive operation. With motors, the starting torque will be reduced and overheating may occur if the motor is fully loaded.

Electrical equipment is normally designed to work for a supply voltage variation of +/- 6%, and the lower voltage limit can be extended to -15% for systems where just resistive loads and fluorescent lamps are used. Below this voltage incandescent lamps will operate very inefficiently and starting problems will occur with fluorescent lamps.

Operating frequency

Generator considerations

Increasing the operating frequency will reduce the excitation current required to achieve rated voltage, as shown in Figure 12. This results in improvements in efficiency and maximum power output, as shown in Appendix 2.

Load considerations

The operation of some domestic electrical appliances, such as television sets, record players and clocks was, at one time, dependent upon their synchronization to the supply frequency, and they therefore required a well regulated supply. This is not the case with more modern equipment since an internally generated reference frequency is used. Nowadays, the limits of acceptable frequency variation are principally set by the requirements of transformers and motors.

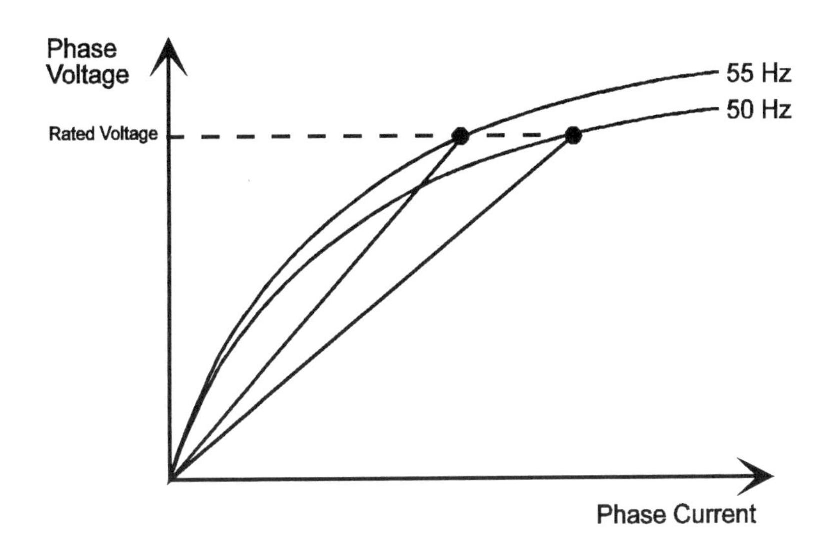

Figure 12 No load excitation characteristics for an induction machine operated at two different frequencies

The effects of supply frequency on motor performance depends on the type of motor used. Only two types of motor are commonly used on a.c. supplies; these are universal motors and induction motors. The universal motor is so called because it gives comparable performance on both d.c. and a.c. supplies. Conversely, induction motor performance is greatly affected by the supply frequency and as a result these machines provide the main limits to its value.

The magnetizing current for an induction motor increases if the supply frequency falls, in the same way as for an induction generator. This increases the power dissipation in the stator windings and can cause them to overheat. This same effect occurs with transformers. Small transformers, which are often used in electrical appliances, are very susceptible to damage from such under frequency operation. For these reasons, **operation at below rated frequency should be avoided**.

Increasing the operating frequency has the opposite affect. It reduces the magnetizing current and therefore induction motors and transformers run cooler. However, there are disadvantages. The main constraint occurs if motor loads, such as pumps and fans, with a power requirement that increases significantly with speed are used on the supply. The effect of such loads can be appreciable since, for induction motors, shaft speed increases approximately linearly with frequency, and the power requirement of fans increases at nearly the cube of the speed. A 10% increase in frequency is generally acceptable, even with such highly speed dependent loads. The reasons for this are as follows:

1) Because of oversizing, most motors run at only 60 - 80% of their rated output when operated at rated frequency. Hence, there is some spare capacity.
2) As previously explained, higher frequency operation reduces the magnetizing current, thereby offsetting some of the increase in load current.
3) Increased shaft speed increases the cooling of the machine.

If no highly speed dependent motors are used then the upper frequency limit can be increased to 20% above rated frequency. If only resistive loads are used then there is no frequency limit imposed by the load. The upper limit will then be determined by the limit of stable operation of the generator, i.e. the frequency at which the machine is no longer saturated for the required voltage output.

Using the operating voltage and frequency to advantage

If an induction machine is operated as a generator at the same voltage and frequency that it is designed for as a motor it will run at lower efficiency and higher temperature than for an equivalent power output as a motor. This is due to increased saturation and therefore greater magnetizing current. Careful selection of the rated voltage of the induction machine and the operating voltage and frequency can reduce the level of saturation and result in the following advantages:

- Improved generator efficiency
- Increased maximum generator power output (within thermal limits)
- Increased winding life
- reduced capacitor costs

The operating voltage is determined by the voltage rating and voltage requirements of the load and the transmission line voltage drop. Ideally, the induction machine should be operated as a generator at 6% below the motor voltage rating, to give the advantages listed above. For example, to operate at 240 V a 255 V induction motor should be used. Chapter 3 presented a number of options for obtaining a machine with the correct voltage rating. If the desired motor voltage rating cannot be obtained then consider operating the loads at slightly above their rated frequency, as this will also provide the advantages listed above.

Though less efficient, it is quite acceptable to operate the generator at the same voltage and frequency that it is designed for as a motor. More capacitance will be required, compared to the values calculated using the methods given in Chapter 4, and extra care should be taken to ensure that the winding currents are not too high.

6. THE EFFECT OF LOAD UPON GENERATOR OUTPUT

Resistive load

Figure 13 shows the voltage variation with resistive load for an induction generator operating at constant speed and with fixed excitation capacitance. The full load current is the current at which the generator current equals the rated current as a motor. Excessive load will cause excitation to collapse.

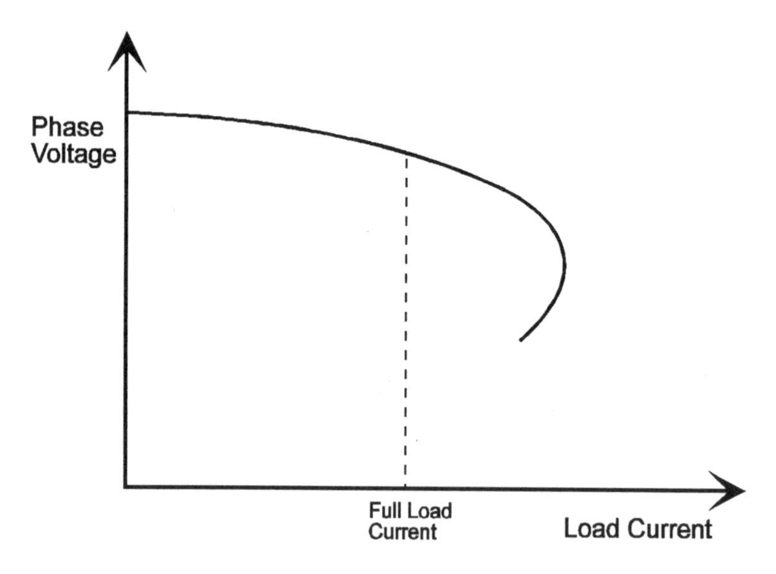

Figure 13 Typical voltage against load current for an induction generator with fixed excitation capacitance operated at constant speed

Figure 13 does not provide a full representation of what occurs when the generator is used on a micro hydro scheme, since it takes no account of frequency variation due to the power-speed characteristic of the turbine. A typical turbine power speed characteristic is shown in Figure 14. When the load on the generator is increased the voltage will drop immediately by an amount determined by the characteristic of

Figure 13. The increased load on the turbine will cause the turbine speed to drop, resulting in a corresponding reduction in operating frequency. This drop in frequency causes a further drop in voltage, due to reduced excitation, as shown in Figure 12. The turbine speed and generated voltage will fall until a speed is reached at which the power output of the turbine equals the load on the turbine.

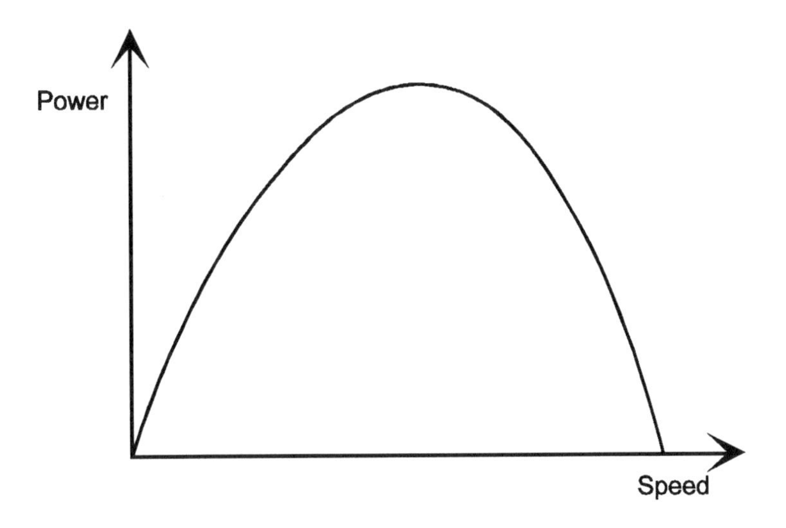

Figure 14 Typical power speed characteristic for an impulse turbine operated at constant head and flow

Conversely, if the resistive load on the generator is reduced the generated voltage will rise and the turbine speed will increase. If all the load on the generator is disconnected the voltage and turbine speed will increase until the losses in the generator equal the power output from the turbine.

It is clear from the above that the resistive load must be controlled in order to regulate the voltage and frequency. Load control options are presented in Chapter 9.

Inductive load

If an inductor is connected across the output of the generator the current into the inductor will be supplied by the capacitors connected to the generator. This will reduce the amount of excitation capacitance available, and as a result the voltage will fall temporarily. The drop in voltage will reduce the power dissipated in the other loads and therefore the turbine speed will increase. The voltage will increase as the speed increases until equilibrium is reached at a speed where the power output of the turbine equals the load on the turbine. Hence, inductive loads result in an increase in operating frequency.

In an uncontrolled system the new operating voltage will depend largely upon the power-speed characteristic of the turbine. If the turbine power output does not vary appreciably between the initial and final speeds there will be no significant change in voltage.

If too large an inductive load is connected the excitation of the generator will collapse. This can occur when connecting induction motors as loads, as discussed in Chapter 10.

Capacitive load

In the same way that an inductive load will increase the operating frequency, a capacitive load will reduce it. This could be damaging to the generator and other loads, as explained in Chapter 5. Fortunately, leading power factor loads are very rare. They may occur if too much power factor correction capacitance is connected across an inductive load.

7. SINGLE-PHASE OUTPUT FROM A THREE-PHASE MACHINE

Single-phase induction motors can be used as generators, but problems can be experienced in achieving excitation and in determining the size and arrangement of the capacitors required. In addition, single-phase induction motors are more expensive than three-phase induction motors and are only available for small power outputs. Fortunately, it is possible to use a three-phase induction motor as a single-phase generator and this is the preferred approach to providing a single-phase supply. The method for obtaining a single-phase output from a three-phase machine is as follows:

1) Use a three-phase induction machine suitable for 240/415 V operation and connect the machine in delta.
2) Calculate the per phase capacitance, 'C', required for normal three-phase 240 V delta operation.
3) Instead of connecting 'C' to each phase connect twice 'C' to one phase, 'C' to a second phase and no capacitance to the third phase. This is known as the 'C-2C' connection. The load should be connected across the 'C' phase, as shown in Figure 15.

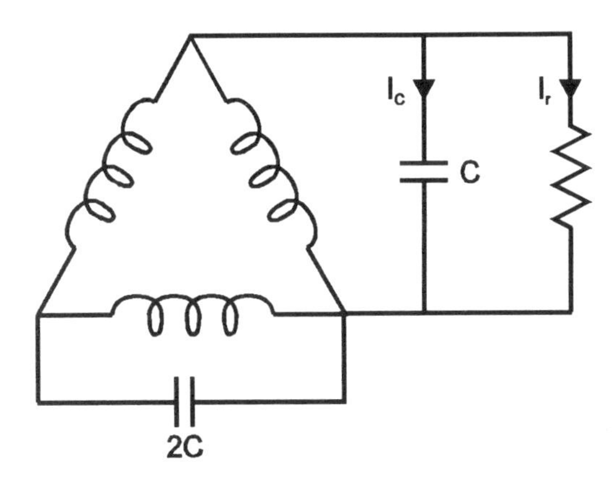

Figure 15 Single phase generation from a three phase machine using the 'C-2C' connection

This unbalanced arrangement of capacitance helps to compensate for the unbalanced load on the generator, and as a result the generator can be used with an output of up to 80% of the motor rating. This is the same derating factor that is applied to three-phase machines to compensate for load imbalance.

In using the 'C-2C' connection it is essential to ensure that the direction of rotation of the machine rotor is correct, in relation to the phases to which the 'C' and '2C' values of capacitance are connected. If the capacitors are arranged correctly the 'C phase' will produce its peak voltage before the '2C phase'. If the opposite occurs the generator will run inefficiently and overheat.

It is possible to determine the correct rotation from the labelling of the windings. However, it is quite easy to make mistakes or to find that the windings have been incorrectly labelled. Correct rotation should be checked by running the machine with both phase sequences. This is a relatively straightforward task. The correct rotation can be determined by measuring the power output or the winding currents.

Power output test

For this test a constant input power is required. This is usually achieved by running the turbine at maximum power output. The turbine output is gradually increased and load applied to the 'C phase' to maintain the voltage at its rated value. When maximum turbine power is reached, measure the load current. Repeat the test with the '2C' capacitance connected to the phase that previously had no capacitance across it. A higher load current reading will be obtained for the correct capacitance arrangement because the generator will run more efficiently. A 10% increase in electrical output is typical. The efficiency improvement will mean that much less power will be dissipated in the generator and as a result it will run appreciably cooler.

This test is very easy to perform if an induction generator controller is used because the controller will maintain a constant voltage and the ballast meter will indicate which connection produces the maximum power output.

Winding currents test

By detailed analysis, it can be shown that for a purely resistive load, if the condition given in Equation 7 is met, the 'C-2C' connected generator behaves as a balanced three-phase machine. Hence, if the winding currents are measured for this load condition the currents will be equal, provided that the phase sequence is correct.

$$P_{load} = \frac{\Sigma Q}{\sqrt{3}} \tag{7}$$

Where P_{load} is the power dissipated in the load

ΣQ is the total reactive power from the capacitors

An alternative representation, using the symbols in Figure 15 is,

$$I_r = \sqrt{3}\, I_C \tag{8}$$

For loads above and below the load condition given in Equation 7 the machine is unbalanced, and as a result the machine will run hotter and less efficiently than a balanced three-phase machine, as shown in Appendix 2. Close to the balanced condition this effect is quite small and derating to 80% of the motor output is more than sufficient compensation for the imbalance.

The balance condition given in Equation 7 is for a resistive load. With the 'C-2C' single-phase system, great care should be taken to power factor correct inductive loads, so as to keep the generator operating as close to its balanced condition as possible.

Example 6

For the 2.2 kW machine of Example 3, calculate the generator power output required to give balanced operation with the 'C-2C' connection.

From Example 3, $\Sigma Q = 2516\,\text{VAR}$

For balanced operation, Equation 7 must be satisfied,

$$P_{load} = \frac{\Sigma Q}{\sqrt{3}} = \frac{2516}{\sqrt{3}} = \underline{\underline{1453\,\text{W}}}$$

8. GENERATOR PROTECTION

Overload

Provided that the generator rating is greater than or equal to the maximum electrical output from the turbine-generator, an overload of the generator windings, due to excess load on all phases, will not be possible. The reason for this is that the partial collapse of excitation caused by the overload will limit the winding currents to safe values. However, with a three-phase system a severe overload on one phase could cause that phase to be burnt out. Miniature circuit breakers (MCBs) should be used to prevent this. MCBs rated for 1.5 times the rated full load generator output can be used, since the machine output has been derated to allow for some load imbalance. If three-phase motor loads are being used then a three-pole MCB should be used to prevent single phasing of the motor if the trip operates.

If the generator rating is less than the maximum electrical output from the turbine-generator, MCBs should be fitted to the generator output to prevent damage due to overloading. **Note** that irrespective of whether a three-phase or single-phase 'C-2C' system is used the maximum generator rating is 80% of the motor rating.

Underload

The induction generator is at greater risk from loss of load than from too much load. The reason for this is that, with little or no load connected, the generator will speed up and hence the frequency and voltage will increase. Both of these factors increase the current flowing into the generator from the capacitors and hence the winding current will increase, generally to a level that exceeds its rated value. To protect against damage to the capacitors and the windings, MCBs should be fitted in series with the capacitors in order to switch them out if the generator overspeeds. Such an arrangement is shown in Figure 16. The current rating of the MCBs should be slightly above the maximum capacitor current under normal operating conditions. It is preferable to disconnect all the capacitors at once. Hence, a three-pole MCB should be used with a three-phase system and a two-pole MCB with a 'C-2C' single-phase system. The two-pole MCB should be rated for the maximum current through the '2C' set of capacitors. Single-pole MCBs can be used if two and three-pole versions are unavailable.

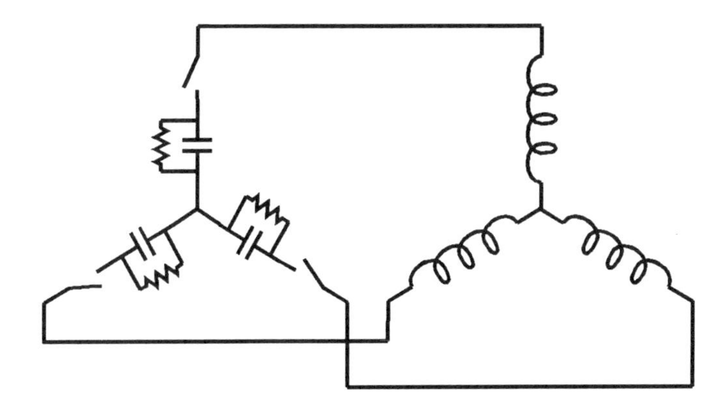

Figure 16 Underload protection by means of MCBs.

Magnetic MCBs are preferable to thermal MCBs as they are more accurate and unaffected by temperature variations. The MCBs should be rated for between 1.2 and 1.5 times the capacitor current under normal conditions. If the MCB has too high a current rating then there is a danger that it will not trip when all the load is removed. Standard current ratings for MCBs are 2, 4, 6, 10, 16, 20, 25, 32, 40 and 50 Amps. The current setting of some MCBs can be manually adjusted. If MCBs with a suitable current rating cannot be obtained then a combination must be used. For example, if a three-phase generating system has a capacitor current at rated frequency and voltage of 10 Amps per phase the MCBs should be rated for between 12 and 15 Amps. Since there is no suitable MCB within this range the capacitors in each phase should be split into two equal sets and protected by 6 Amp MCBs.

The correct operation of the MCBs should be checked by performing a test at runaway speed when the generator is installed. This is done by disconnecting the consumer load and any ballast load, and opening the turbine valve fully. The power output of the turbine will be dissipated in the generator and drive. If the MCBs fail to trip within ten minutes then they must be replaced with MCBs of a lower current rating or greater sensitivity. For small overcurrents MCBs take several minutes to operate.

Small generating systems tend to have the lowest runaway voltage, because, as shown in Table 1, the induction machine is less efficient and will therefore dissipate the turbine power output at a lower voltage. The runaway voltage is typically twice rated voltage, though for small generators it can be as low as 360 V and for large generators as high as 600 V.

If the turbine operating at runaway speed does not cause the excitation current to exceed the rated current of the generator then MCBs are not necessary, provided that the capacitor voltage rating is not exceeded. This is only likely for turbines with an efficiency that falls off rapidly with overspeed.

If an induction generator controller is used the small amount of capacitance that this contains could be sufficient to cause excitation to persist, even once the main capacitors have been disconnected by the MCBs. This is acceptable since the excitation current under these conditions will be considerably below the rated current of the generator.

Normally when the turbine is closed down the capacitors will discharge into the generator windings. However, if the capacitors are disconnected by the MCBs they will remain charged and could cause an electric shock if touched. The resistors shown in Figure 16 are incorporated for electrical safety as they will ensure that the capacitors are discharged when the turbine is closed down. For the sake of high reliability, ensure that under normal operating conditions the power dissipation in the discharge resistors is less than 25% of their rated power dissipation.

Overspeed

Standard induction machines can withstand considerable overspeeds due to their solid rotors. Twice rated speed is normally quite acceptable, though for 2 pole machines and large machines manufacturer's advice should be obtained. Most micro-hydro schemes use 4 pole generators, with turbines having a runaway speed less than twice rated speed, and therefore the induction generator needs no overspeed protection.

9. FIXED AND VARIABLE LOAD SYSTEMS

Fixed load systems

The simplest induction generator systems have a constant turbine power output and a constant load. No form of load control is required and therefore the cost and complexity of the system is kept to a minimum.

The system is operated by bringing the turbine up to speed and then switching in the electrical load. The turbine power output is then increased until the required voltage is achieved. An overvoltage trip should be used to prevent loads from being damaged if some of the electrical load becomes disconnected. The design for a suitable trip is given in Appendix 5.

The fixed load approach has a number of disadvantages. These are as follows:-

1) It is often difficult to ensure a constant electrical load. For example, on a village micro hydro scheme it might be assumed that a load consisting of light bulbs only, with no individual switches, represents an ideal fixed load. Experience in Nepal has shown that this is not always the case, as villagers may remove the bulbs from their sockets if they decide to go to bed early.
2) Tripping of the system due to overvoltages can be very inconvenient, especially if the generator is far from the load centre. As a result the normal operating voltage may be set at a low level to reduce the chances of tripping. This results in inefficient operation of some domestic appliances, particularly light bulbs.
3) Frequent tripping of the system may result in the overvoltage trip being bypassed by the users.
4) Switchable or variable loads must be avoided. If used they must only make up a small proportion of the total load. For example, a refrigerator can generally be used on a system where there is a large fixed load.
5) In general, less productive use is made of the power available.

To maximise the use of fixed load systems, changeover switches can be installed allowing a choice between two or more appliances of approximately the same power rating. However, a system with an overvoltage trip and changeover switches is less versatile than one with a load controller and may be almost as expensive.

Variable load systems

For a small increase in overall system costs, an electronic controller can be used with the generating system to allow large changes in consumer load to be accommodated. This overcomes the disadvantages for fixed load systems that were listed in the previous section.

The generated voltage and frequency are controlled by maintaining a near constant load on the turbine, as shown in Figure 17. The controller compensates for variations in the main load by automatically varying the amount of power dissipated in a resistive load, generally known as the 'ballast load', in order to keep the total load constant.

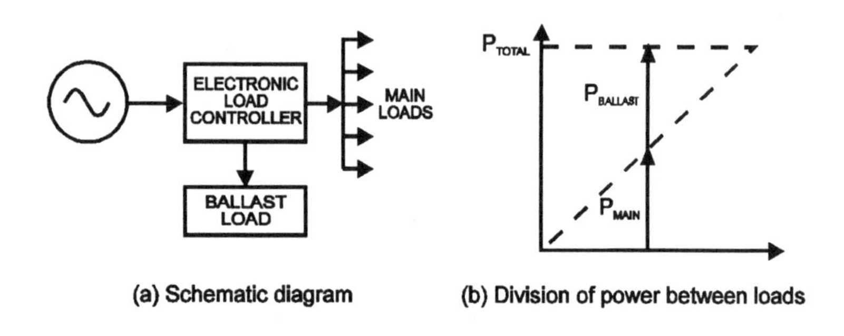

Figure 17 Basic principles of electronic load control

The first electronic load controllers were designed for use with synchronous generators. They sense and regulate the generated frequency. Voltage control is not required because synchronous generators have inbuilt voltage regulators. Without an electronic frequency controller, the frequency will vary as the load changes and under no load conditions will be much higher than rated frequency.

The control requirements of the induction generator are different from those of a synchronous generator, since, without any electronic control, both the voltage and frequency will vary when the load changes. As a result, early control systems contained a voltage controller as well as a frequency controller. The voltage controller controlled excitation by varying the amount of capacitance connected and the frequency controller varied the resistive load, just like the controllers for synchronous generators. Such systems are expensive and complex and therefore have not been widely used.

The induction generator controller (IGC), developed by the author for ITDG, dispenses with the need for a separate voltage controller by using the characteristics of the generator and water turbine to advantage. The controller is very similar to the electronic controllers used with synchronous generators, except that voltage is sensed and directly controlled instead of frequency. The characteristics of the generator are such that this controller also produces good frequency regulation.

When a resistive load is connected to the generator the voltage falls. The IGC senses the drop in voltage and compensates by decreasing the power dissipated in the ballast load. A small transient change in frequency will occur before both the frequency and voltage return to their original values. The total load on the generator will be the same as before the load change.

If an inductive load is connected to the supply the voltage will fall, and hence the controller will reduce the power dissipated in the ballast. This reduction in resistive loading causes the turbine to speed up, increasing the generated frequency. The increased frequency reduces the magnetizing current required by the generator and increases the capacitor current. As a result, the voltage will rise and return to its initial value. The frequency will have increased in order to compensate for the inductive load. As a result of the high level of saturation of modern induction machines the frequency increase with inductive loads is small, typically less than 10% for an overall load power factor of 0.9. Given that the ballast load is resistive, it is not difficult to maintain an overall load power factor greater than 0.9. Improved frequency regulation can be obtained by power factor correcting significant inductive loads. If a large number of fluorescent lamps are used on the system this will constitute a significant inductive load.

Controller options

There are a number of options for varying the ballast load: phase angle control, binary weighted loads and mark-space ratio control.

Phase angle control

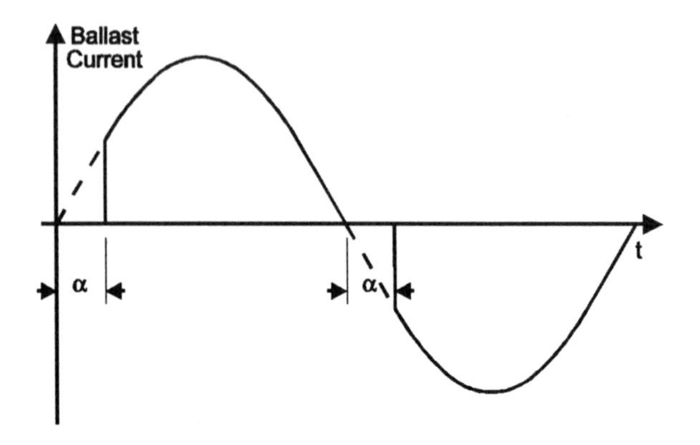

Figure 18 Ballast current waveform for a phase angle controller

By delaying the switching on of a triac or thyristor arrangement a variable resistive load can be produced, as shown in Figure 18. Each half cycle, the switching on of the ballast load is delayed by the phase angle α which can have a value of between 0° and 180°.

This control approach is often used with synchronous generators, but is less appropriate for induction generators because of the variable lagging power factor produced due to the ballast current lagging behind the voltage. The effect of this with an induction generator is to increase the frequency variation that occurs, as explained in Chapter 6. A further disadvantage is waveform distortion which produces increased heating in the generator windings. To compensate for the waveform distortion, the generator should be oversized.

Binary weighted loads
In this case the variable resistive load is produced by switching in a combination of fixed resistors. The values of these fixed resistors are binary weighted so as to achieve the maximum number of load steps with the minimum number of resistors and switches. A single-phase, three resistor arrangement is shown in Figure 19, along with the seven values of ballast load that can be achieved.

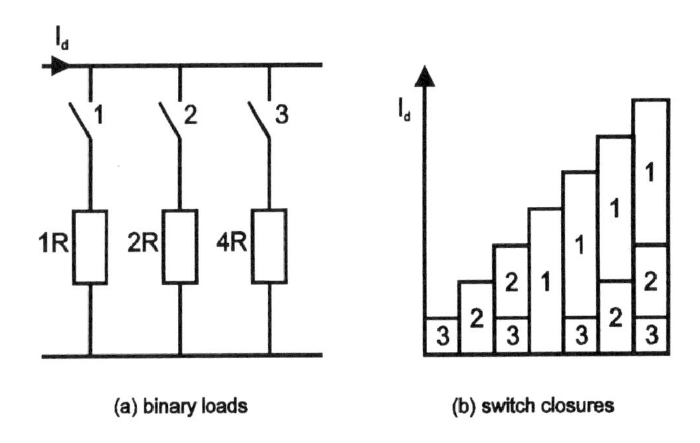

Figure 19 Binary-weighted load controller and ballast current range

The main advantages of this approach are that waveform distortion is not produced and the ballast load is resistive. Binary weighted controllers have been used in Nepal. However, they do have a number of disadvantages. The main disadvantage is the complexity resulting from requiring a number of ballast loads, each with its connections, wires and switching device. The ballasts must be of accurate resistance value, which may prove difficult with available heating elements, especially for small power ratings. In addition, because the ballast load is only varied by fixed steps, the voltage is only controlled within a range or 'window'. For stable operation under part flow conditions a large window must be used, resulting in poor voltage regulation.

Mark-space ratio controller

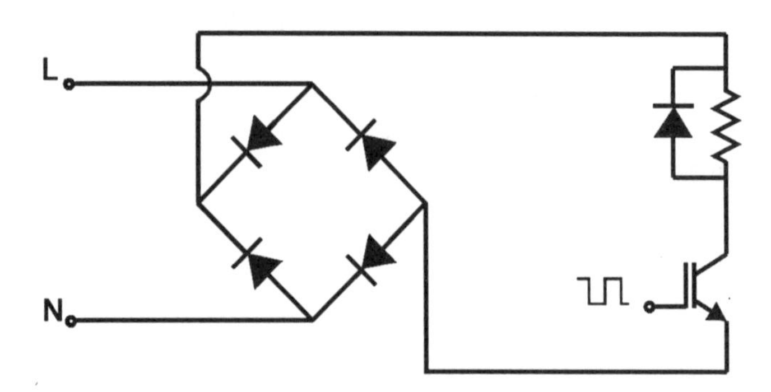

Figure 20 Basic switching circuit for a single-phase mark-space ratio controller

The mark-space ratio controller, in its simplest form, requires just a single ballast load, as shown in Figure 20. The ballast load is connected across the rectified output of the generator and switched on and off by means of a transistor. The controller varies the on (mark) and off (space) times, known as the mark-space ratio, in order to control the power dissipation in the ballast load. The on time can be varied over the full range of 0% to 100%. Figure 21 shows voltage waveforms with 60% on-time.

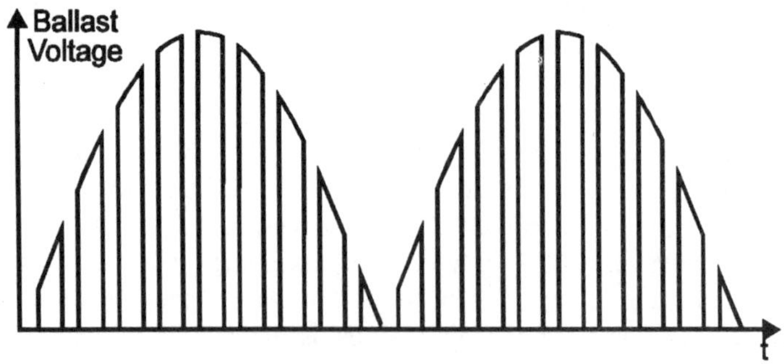

Figure 21 Typical ballast voltage waveform for a single-phase mark-space ratio controller with 60% on time

Figure 22 5 kW locally manufactured mark-space ratio IGC, Agha Khola, Nepal

The advantages of this type of controller are good voltage regulation, simple connection of ballast loads and an effectively resistive ballast load. It is the preferred IGC design and has now been installed on a total of more than forty sites in Indonesia, Nepal, Sri Lanka and England. Figure 22 shows an IGC locally assembled and installed in Nepal.

The three phase version uses a three phase bridge rectifier as shown in Figure 23. Its disadvantages are that there is no phase balancing and there is increased waveform distortion. The advantages are that it is as simple as the single phase design and uses the same circuit board.

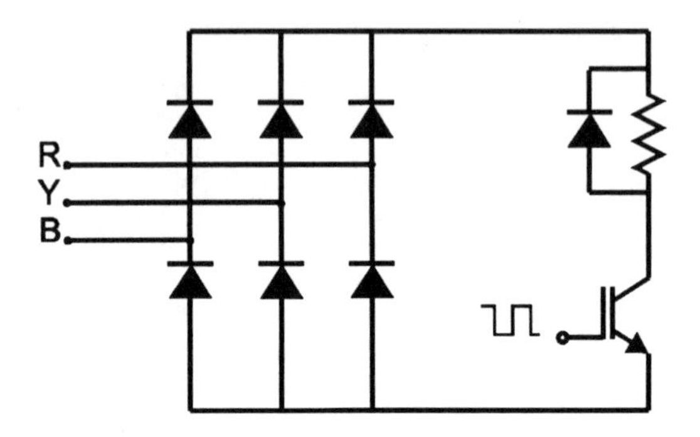

Figure 23 Three phase mark-space ratio controller.

The IGC has been developed for local manufacture in developing countries. A good knowledge of electronics is required in order to manufacture, install and repair it. To ensure good quality local manufacture the IGC design is only provided to engineers who attend an ITDG approved training course.

10. MOTOR STARTING

Motor starting can be a considerable problem with induction generators. The reasons for this are that the starting current for an induction motor is typically 6 times the rated current and the power factor at starting is lower than when the motor is running. For direct-on-line starting the limit for the motor, as a proportion of the generator rating, is typically 15% for a single-phase motor and 10% for a three-phase motor. Single-phase motors are generally easier to start because their starting current is not as high and their power factor is better. If too large a motor is switched on there will be insufficient capacitance to both supply the generator and power factor correct the motor, and the generator will de-excite.

These values are just guidelines, since a lot depends on the load that the motor is driving. Loads requiring a high starting torque, such as compressors, are much harder to start than low starting torque loads such as fans. For example, a 1.1 kW single-phase fan could be started by a 4 kW induction machine connected 'C-2C' and producing 2 kW. Whereas, a 1.1 kW induction machine connected 'C-2C' and producing 800 W was unable to start a 180 W refrigerator compressor.

Techniques for improving motor starting

One option for improving motor starting is to oversize the generator, since this provides extra capacitance that will help to power factor correct the motor at starting. The disadvantages of this approach are increased cost and reduced efficiency, especially at part load. Better options are power factor correction at starting or, for three-phase machines, star-delta starting.

Power factor correction at starting involves temporarily connecting capacitors to compensate for the inductive current. The starting current can usually be obtained from manufacturers' data sheets. It is generally called the locked rotor current, and refers to the initial starting condition when the rotor is stationary. The power factor at starting is low, typically 0.4. Hence, at starting the inductive current is only slightly less than the locked rotor current.

Example 7

The manufacturers' data for a 380 V, three-phase 5.5 kW induction motor are as follows:

Full load current, I_{fl} 12 Amps
Locked rotor current, I_{lr} 70 Amps
Full load power factor, pf 0.89

Calculate the capacitance required to fully power factor correct the motor when it is run at full load.

The total apparent power at full load,

$$\Sigma S = \sqrt{3} \times V_{line} \times I_{fl} = \sqrt{3} \times 380 \times 12 = 7898\,\text{VA}$$

Hence, the real power,

$$\Sigma P = \Sigma S \cos\phi = 7898 \times 0.89 = 7029\,\text{W}$$

The reactive power can be obtained from the power triangle:

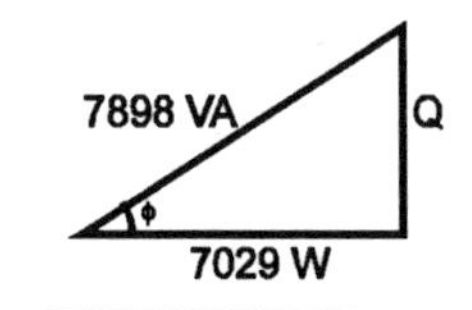

$$\Sigma Q = \sqrt{7898^2 - 7029^2} = 3602\,\text{VAR}$$

$$Q_{phase} = \frac{\Sigma Q}{3} = \frac{3602}{3} = 1201\,\text{VAR}$$

A delta connection of capacitors will normally be chosen as it is cheapest,

$$V_{phase} = V_{line}$$

$$I_{phase} = \frac{Q_{phase}}{V_{phase}} = \frac{1201}{380} = 3.16\,\text{A}$$

From Equation 6, $C = \frac{I}{2\pi f V} = \frac{3.16}{2\pi.50.380} = \underline{\underline{26\,\mu\text{F}}}$

Example 8

For the 5.5 kW motor described in Example 7 the locked rotor power factor is 0.4. Calculate:

(a) The power drawn from the supply for a direct on line start.
(b) The additional capacitance required for full power factor correction at starting, assuming that the motor is already compensated for full load running.

(a)

$$\Sigma S = \sqrt{3} \times V_{line} \times I_{lr} = \sqrt{3} \times 380 \times 70 = 46.07\,kVA$$

$$\Sigma P = \Sigma S \cos\phi = 46.07 \times 0.4 = \underline{\underline{18.4\,kW}}$$

(b)

$$\Sigma Q = \sqrt{\Sigma S^2 - \Sigma P^2} = \sqrt{46.07^2 - 18.43^2} = 42.2\,kVAR$$

$$Q_{phase} = \frac{\Sigma Q}{3} = \frac{42.2}{3} = 14.1kVAR$$

$$V_{phase} = V_{line}$$

$$I_{phase} = \frac{Q_{phase}}{V_{phase}} = \frac{14100}{380} = 37\,A$$

$$\text{From Equation 6, } C = \frac{I}{2\pi fV} = \frac{37}{2\pi.50.380} = \underline{\underline{310\,\mu F}}$$

Subtracting the 27μF connected for full load power factor correction, the additional capacitance required is,

$$C = 310 - 27 = \underline{\underline{283\mu F}}$$

The additional capacitance required at starting is more than ten times the capacitance required when running.

The capacitance required to power factor correct at starting is much higher than that required when running, as shown by Examples 7 and 8. If standard capacitors are used then the cost of the power factor correction will be very high. However, since the capacitors need only be connected for a few seconds, whilst the motor runs up to speed, capacitors rated for intermittent operation can be used. 'Motor start' electrolytic capacitors are the obvious choice since they are designed to be operated for the short time required for motor starting. Their cost per micro-Farad is typically between 10% and 20% of the cost of 'motor run' capacitors.

For switching the capacitors it is best to use solid-state relays (SSRs) of the type that switch at zero current to reduce voltage transients. Contactors are not recommended because the arcing that occurs at switching will rapidly wear them out. A control circuit for a single-phase system is given in Appendix 5. With three-phase systems star-delta starters are a cheaper and simpler option provided that the motor is rated 415 V when delta connected and the load is such that the motor will start with 1/3 of the direct-on-line starting torque.

With star-delta starting the motor is initially connected in star and once up to speed it is reconnected in delta. This is done manually or automatically depending on the type of starter used. The two winding configurations are shown in Figure 24.

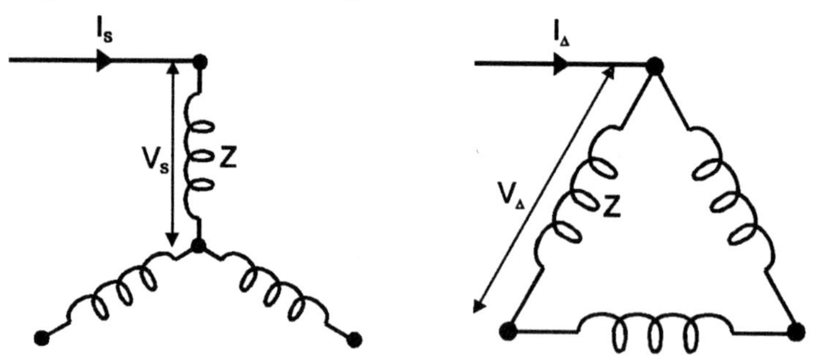

Figure 24 Star and delta starting conditions

At starting, the per phase impedance, Z, is the same whether the motor is connected in star or delta. Hence,

$$V_S = I_S.Z, \quad V_\Delta = \sqrt{3}.V_S$$

$$I_\Delta = \sqrt{3}\frac{V_\Delta}{Z} = \sqrt{3}\frac{\sqrt{3}.V_S}{Z}$$

Hence, $$I_\Delta = \frac{\sqrt{3}.\left(\sqrt{3}.I_S.Z\right)}{Z} = \underline{\underline{3.I_S}} \qquad (9)$$

Since the line current is reduced by a factor of three for a star connection, The real power and reactive power will be reduced by a factor of three at starting.

Example 9 shows that by using a star-delta starter the starting capacitance can be reduced by a factor of more than three. As a result, it is usually possible to start motors with a rating of one third of the capacity of the generator. For example, at an installation in Indonesia with an 18.5 kW induction motor running as a generator, and an output of 10 kW, it was possible to start a saw mill motor of 7.5 kW. To achieve this a star-delta starter was used and the motor was power factor corrected for the running condition.

Example 9

Repeat Example 8(b) for star-delta starting.

The reactive power is reduced by a factor of three. Hence,

$$I_{phase} = \frac{Q_{phase}}{V_{phase}} = \frac{14070/3}{380} = 12.3\,A$$

From Equation 6, $$C = \frac{I}{2\pi fV} = \frac{12.3}{2\pi.50.380} = \underline{\underline{103\,\mu F}}$$

Subtracting the 27μF connected for full load power factor correction, the additional capacitance required is,

$$C = 103\text{-}27 = \underline{\underline{76\mu F}}$$

This is just 27% of the additional capacitance required for the direct on line case.

11. GENERATOR COMMISSIONING

Safety

Wiring and earthing should be carried out to national standards and tested thoroughly. Mechanical safety, including provision of adequate guards, should be checked before starting the turbine.

Output frequency and power

When the generator is commissioned, adjust the turbine power output in order to obtain rated voltage. Then, whilst maintaining a constant voltage, vary the load between no load and maximum generator output and measure the range of generated frequency. Check that the frequency range is acceptable (see Chapter 5 for details) and if not change the amount of excitation capacitance connected. If inductive loads are to be used these must be connected in order to determine whether the increase in generated frequency is acceptable.

Check the winding currents at maximum load to ensure that the current rating of the induction machine is not exceeded. Due to inbalance with the 'C-2C' connection it is acceptable for one of the windings to be overloaded, provided that the other two are underloaded. The following condition must be met:

$$I_1^2 + I_2^2 + I_3^2 \leq 3 \times I_{rated}^2 \quad (10)$$

Where I_1, I_2, I_3 are the three measured winding currents

I_{rated} is the rated winding current

If the generator rating is less than the maximum electrical output from the turbine generator, either fit a larger generator or ensure that there is adequate overload protection, as explained in Chapter 8.

Phase rotation

If a three-phase system is installed the phase sequence should be checked to ensure correct rotation of any three-phase motor loads that are used.

If a single-phase 'C-2C' system is used it is essential that the correct phase sequence is used to prevent the generator from overheating, as explained in Chapter 7.

Underload testing

The testing of underload protection should be carried out, as explained in Chapter 8.

Adjustment for maximum efficiency

The system efficiency can sometimes be improved by adjustment of the amount of capacitance connected. The resulting change in speed will cause a change in both turbine and generator efficiency and, in the case of a reaction turbine, the penstock losses. The frequency must be kept within the acceptable limits given in Chapter 5.

APPENDIX 1: Three-phase circuit theory

This section presents a summary of the essential terms and relationships used in the main text, along with worked examples. For more comprehensive coverage refer to electrical engineering textbooks that cover three-phase circuits.

Line and phase quantities

The terms *line* and *phase* are applied to voltages and currents when describing three-phase circuits, and therefore it is important to be clear about their meaning.

In a single-phase circuit, the two wires connecting the load are called the *line*. However, if one of the wires is earthed, the unearthed wire is usually referred to as the *line* to distinguish it from the earthed one, which is called the neutral. In a three-phase four-wire system the *line* consists of four wires; three *lines* (L_1, L_2, L_3) and the neutral (N), as shown in Figure A1(a). The voltage between any two *lines* is referred to as the *line-to-line* voltage, but often shortened to the *line* voltage. If the load is balanced (equal on each phase) the current in the neutral conductor will be zero and this conductor can be dispensed with, thereby making a three-wire system. An alternative three-phase three-wire system, again consisting of just three *lines* (L_1,L_2,L_3) is shown in Figure A1(b).

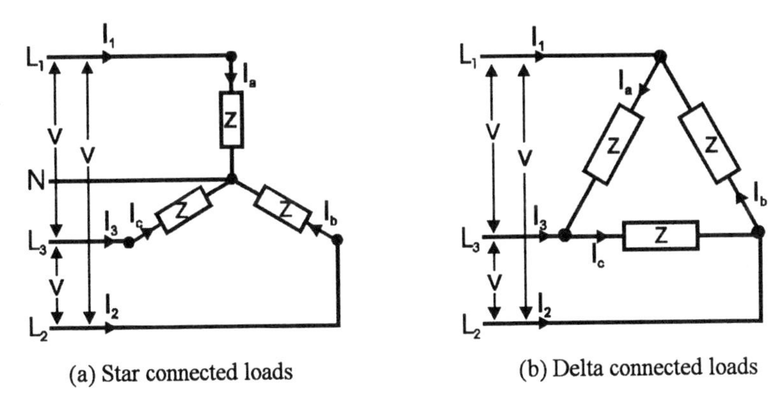

Figure A1 Voltage and current relationships in three-phase circuits

A *phase* is one of the three branch circuits making up a three-phase circuit. In a star connection, a *phase* consists of those circuit elements connected between one *line* and neutral. In a delta connected circuit, a *phase* consists of those circuit elements connected between two *lines*. Hence, in a star circuit the *phase* voltage is the voltage between *line* and neutral and in a delta circuit it is the voltage between two *lines*. The *phase* current is the current that flows through the circuit elements that make up the *phase*.

Three-phase supplies and loads are specified in terms of their *line-to-line* voltage and *line* current. Hence, motor data, such as that given in Table 1, is specified using *line* values. Since this is the standard representation for three-phase supplies, there may be no explicit reference to indicate that *line* values are being used.

Voltage and current relationships

It is clear from Figure A1 that for a star connected system the line currents (I_1,I_2,I_3) equal the phase currents (I_a,I_b,I_c), and that for a delta connected system the line voltages equal the phase voltages. By means of phasor diagrams, it can be shown that for a star connected system the line voltage is √3 times the phase voltage and for a delta connected system the line current is √3 times the phase current. These relationships are summarised in Table A1.

Quantity	Star connection	Delta connection
Voltage	$V_{line} = \sqrt{3}\ V_{phase}$	$V_{line} = V_{phase}$
Current	$I_{line} = I_{phase}$	$I_{line} = \sqrt{3}\ I_{phase}$

Table A1 Voltage and current relationships in three-phase circuits

Power and Power factor

In a.c. circuits the product of the r.m.s. values of the applied voltage and current supplying a load is the apparent power VI *voltamperes*, the latter term being used to distinguish this quantity from the real power, expressed in *watts*. For a resistive load the real power equals the apparent power. For a partially inductive or capacitive load the real power is less than the apparent power, and the latter has to be multiplied by a quantity termed *power factor* to give the power in watts.

$$P = S\cos\phi$$

Where S = Apparent power [VA]
P = Real power [W]
$\cos\phi$ = Power factor

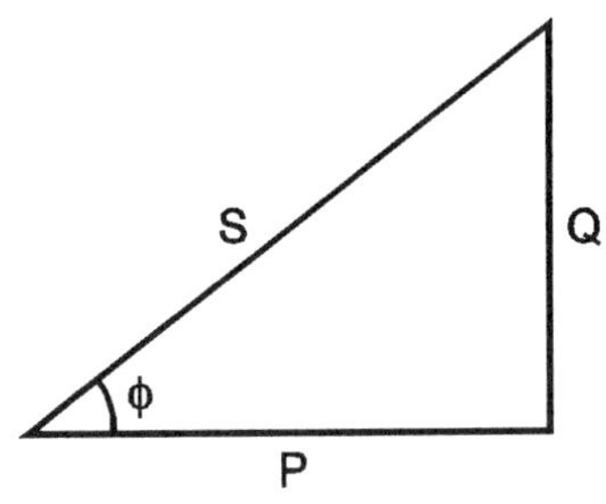

Figure A2 Power triangle

A power triangle can be drawn, as shown in Figure A2. The third side of the triangle represents the reactive power. The relationship between apparent, real and reactive power is:

$$S = \sqrt{P^2 + Q^2}$$

Where Q = Reactive power [VAR]

If we refer to the star connected load of Figure A1(a), the apparent power supplied to each phase is:

$$S = \frac{V_{line}}{\sqrt{3}} \times I_{line} \quad \text{or simply } \frac{V}{\sqrt{3}} \times I$$

The total apparant power supplied to all three phases is therefore:

$$S_{total} = \frac{V}{\sqrt{3}} \times I \times 3 \quad = \sqrt{3}\ VI$$

For a delta connected load the result is the same, since:

$$S_{total} = V \times \frac{I}{\sqrt{3}} \times 3 \quad = \sqrt{3}\ VI$$

Hence, for calculating the power drawn by a motor, by means of manufacturers' data, it is not necessary to know whether it is star or delta connected.

Example A1

A 3 kW, three-phase motor, connected to a 415 V 50 Hz supply, draws a line current of 6.5 A. If the power factor is 0.8, calculate:
(a) The apparent power supplied.
(b) The real power supplied.
(c) The reactive power supplied.
(d) The efficiency, assuming that the power output of the motor is 3 kW.

(a) The apparent power is:

$$S = \sqrt{3}\ VI \ = \sqrt{3} \times 415 \times 6.5 = \underline{\underline{4672\ VA}}$$

(b) The real power is:

$$P = S\cos\phi \ = 4672 \times 0.8 = \underline{\underline{3738\ W}}$$

(c) The reactive power can be obtained from the power triangle:

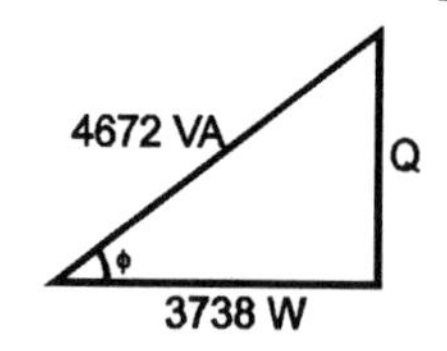

$$Q = \sqrt{4672^2 - 3738^2} \ = \underline{\underline{2803\ VAR}}$$

(d) The efficiency is: $\frac{\text{Mechanical output power}}{\text{Electrical input power}} = \frac{3000\ W}{3738\ W} = \underline{\underline{0.80}}$

For stand-alone induction generators, capacitors must be used to fully power factor correct the induction machine, i.e. increase the power factor to 1.0, by providing sufficient VARs to meet the VAR requirement of the machine. Example A2 illustrates the calculations required.

Example A2

For the motor specified in Example A1, calculate the capacitance required for full load power factor correction:
(a) For star connected capacitors.
(b) Delta connected capacitors.

As calculated in Example A1, the total reactive power required by the motor , Q, is 2803 VAR.

Hence, the reactive power per phase,

$$Q_{phase} = \frac{Q}{3} = \frac{2803}{3} = 934\,\text{VAR}$$

(a) STAR: $$V_{phase} = \frac{V_{line}}{\sqrt{3}} = \frac{415}{\sqrt{3}} = 240\,\text{V}$$

$$I_{phase} = \frac{Q_{phase}}{V_{phase}} = \frac{934}{240} = 3.89\,\text{A}$$

$$X_c = \frac{V}{I} = \frac{1}{2\pi fC}$$

Hence, $$C = \frac{I}{2\pi fV} = \frac{3.89}{2\pi.50.240} = \underline{\underline{52\,\mu\text{F}}}$$

(b) DELTA: $V_{phase} = V_{line} = 415V$

$$I_{phase} = \frac{Q_{phase}}{V_{phase}} = \frac{934}{415} = 2.25A$$

$$C = \frac{I}{2\pi fV} = \frac{2.25}{2\pi .50.415} = \underline{\underline{17\,\mu F}}$$

APPENDIX 2: Induction generator efficiency

The following tables and graphs present efficiency and winding temperature results for a 230 V, 50 Hz, three-phase, 4-pole, 2.2 kW induction motor when operated as a motor and a generator. For generator operation the excitation capacitance was varied as the load was changed in order to maintain a constant voltage at fixed frequency.

Load (% motor rating)	Temperature (°C)	Efficiency %
25	60	71.8
50	65	81.6
75	73	84.1
100	82	83.4
125	91	81.6

Table A2 Motor performance (50 Hz and 230/400 V star connected)

Connection	Load (% motor rating)	Capacitance (μF)	Temperature (°C)	Efficiency %
Balanced Delta	25	47	64	65.6
	50	52	76	74.2
	75	58	82	75.8
	100	66	92	75.9
C-2C	25	43	66	64.3
	50	50	75	73.9
	75	58	86	75.8
	100	67	95	75.6

Table A3 Generator performance for balanced delta and 'C-2C' connections (50 Hz and 230V)

Phase Voltage	Frequency (Hz)	Load (% motor rating)	Capacitance (μF)	Temperature (°C)	Efficiency %
230	50	25	44	59	67.3
230	50	50	48	69	75.4
230	50	75	55	81	78.1
230	50	100	63	91	78.0
215	50	25	37	46	71.1
215	50	50	40	54	78.0
215	50	75	47	68	82.3
215	50	100	55	82	79.8
200	50	25	33	47	70.6
200	50	50	37	56	77.6
200	50	75	43	68	80.5
200	50	100	50	80	78.8
230	55	25	28	52	72.8
230	55	50	31	60	80.1
230	55	75	35	68	81.5
230	55	100	40	76	81.1
230	55	125	46	86	80.4

Table A4 Generator performance for a star connected machine under a range of voltage and frequency conditions.

The tables and graphs clearly show that induction machines are more efficient when operated as motors than as generators. Improvements in generator mode performance can be obtained by operating at less than rated voltage and more than rated frequency. For motor and generator mode operation, maximum efficiency is generally achieved at about 80% of motor rating. Hence, restricting maximum generator mode power output to 80% of motor rating, as well as protecting against excessive winding temperatures, tends to improve efficiency.

With 'C-2C' operation efficiency is always lower than for balanced three-phase delta operation, except at the balance point. However, the difference in efficiency away from the balance point is quite small. Delta operation is less efficient than star operation due to increased harmonic currents.

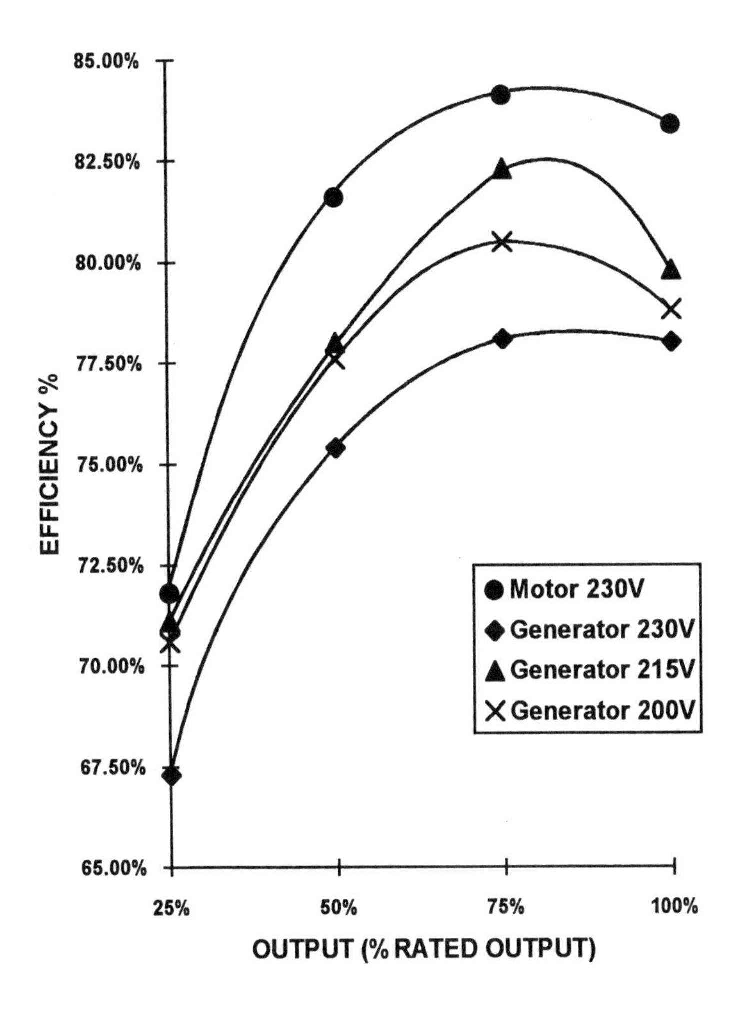

Figure A3 Efficiency curves for generator and motor operation (star connected and 50 Hz)

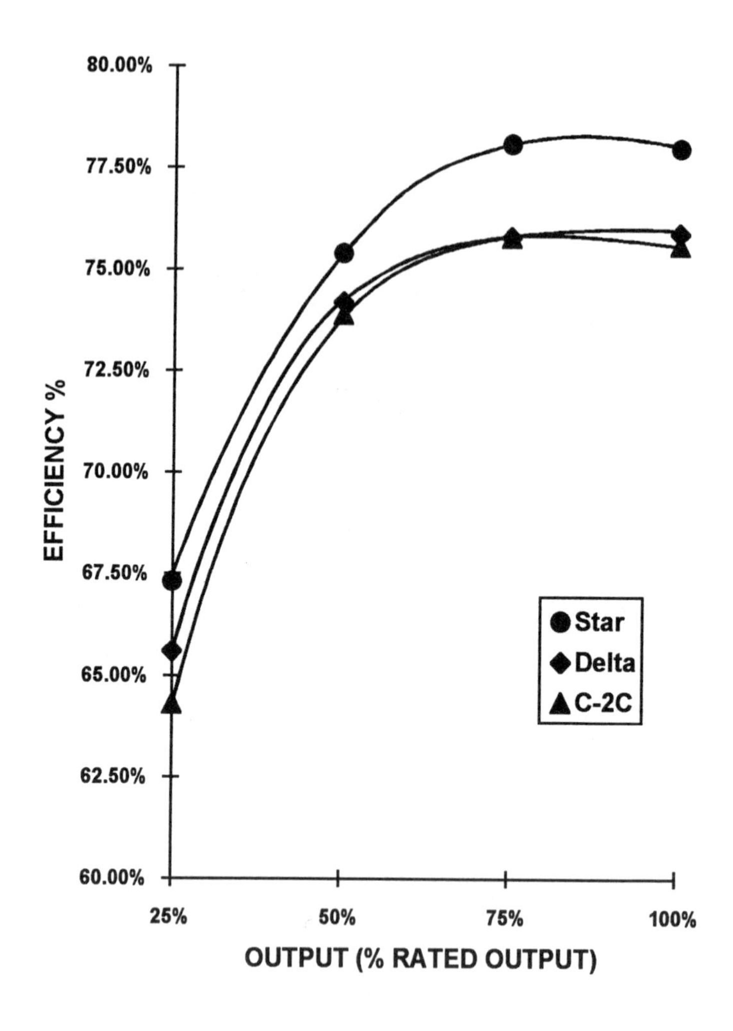

Figure A4 Efficiency curves for generator operation with various configurations (230 V and 50 Hz)

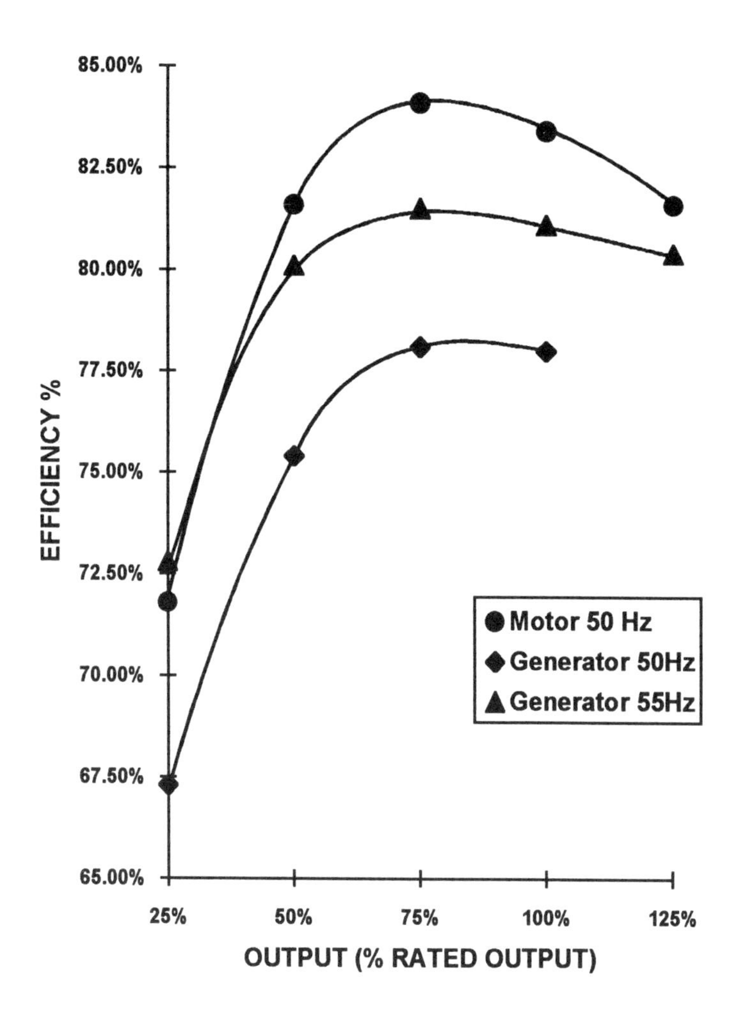

Figure A5 Efficiency curves at 230 V and two frequencies

APPENDIX 3: Single transformer systems

If a long transmission line is required, it is worth stepping the voltage up and down by means of transformers in order to reduce cable costs. However, the savings in terms of cable cost must be compared with the expense and additional losses introduced by the transformers.

An intermediate option is to generate at a high voltage so that only one transformer, the step down transformer, is required. This is particularly useful for single-phase systems, since all standard induction motors can be connected for 415 V line to line and the 'C-2C' connection applied. 415 to 240 V step-down transformers are relatively easy to obtain. Transmitting at 415 V rather than 240 V means that the cable cross-sectional area can be reduced by a factor of three for the same power loss. The transformer represents a significant inductive load and should therefore be power factor corrected.

APPENDIX 4: Halving the operating voltage of an induction machine

In most cases, each phase of a standard induction machine consists of an even number of groups of coils connected in series. The voltage rating can be halved and the current rating doubled by reconnecting the groups into two parallel sets. This is achieved by breaking and then rejoining the connections between the groups of coils that are found in the end windings of the machine. No rewinding is required.

This operation can be carried out by a competent machine winder or any engineer equipped with the necessary heat resistant sleeving, lacing, varnish and brazing equipment. Ideally the windings should be joined by brazing, as this produces a strong connection that is able to withstand vibration. If brazing equipment is not available then the joints can be soldered using a heavy duty soldering iron. With soldered joints extra care should be taken when varnishing and relacing to fix the joints so that they cannot flex.

A series to parallel conversion for a machine with two groups of coils per phase will be described, since this is commonly found with both two and four pole machines. The procedure is as follows:

1) Isolate all six ends of the windings at the terminal block.

2) Expose the connections in the end windings. These are usually at the drive end of the machine and will be covered with heat resistant sleeving.

3) Pull the connections away from the windings, cutting the lacing as required.

4) Cut away the heat resistant sleeving to expose the connections.

5) For a machine with two groups of coils per phase there will usually be three brazed joints, per phase, as shown in Figure A6(a). Leads from the terminal block connect to the ends of the groups (joints A and C in the diagram). The middle connection (joint B) is between the two groups of coils. The three joints can easily be identified using a multimeter with a resistance setting.

6) Label the wires corresponding to U1 and U2 in Figure A6(a). Cut joint B and using the meter identify and label U1' and U2'.

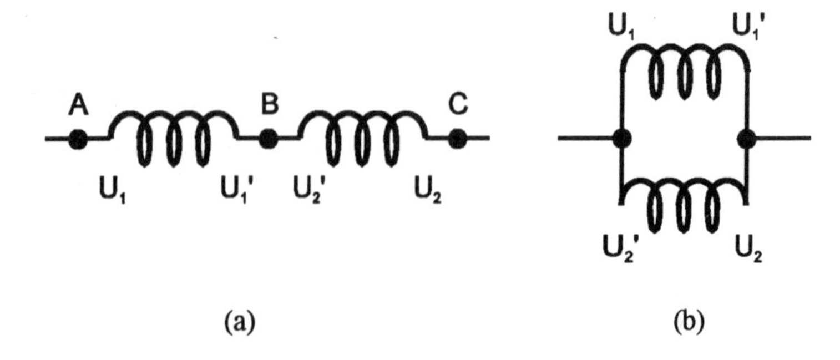

Figure A6 Series to parallel conversion of coil connections

7) Connect U1' to U2 and U2' to U1, as shown in Figure A6(b), covering the connections with heat resistant sleeving. Note that the cables going to the terminal block will now carry twice their original current. If these cables are not sufficient for the higher current then they must be replaced.

8) Repeat steps 5, 6 and 7 for the other two phases.

9) Relace the windings and joints and apply varnish to the heat resistant sleeving and any areas where the old varnish has been disturbed.

10) Test the windings using an insulation tester.

If there are four or more groups of coils per phase, the middle connection (joint B) is at the middle of the series. The same procedure can be used.

APPENDIX 5: Overvoltage trip

A circuit to protect the electrical loads in the event of an overvoltage is shown in Figure A7. This should be used with all fixed load systems, as explained in Chapter 10. When an overvoltage occurs, the relay contacts open and the contactor coil becomes de-energized causing its contacts to open and disconnect the loads from the generator.

The only component that is dependent upon the generator rating is the contactor, all the others are standard. The current rating of the contactor must be sufficient to supply the maximum load on the generator. Two current ratings are usually given for a contactor: an AC1 rating and an AC3 rating. The AC1 rating is for largely resistive loads and the AC3 rating for motor loads. Although, for a fixed load system, the load will fall into the AC1 category, the AC3 rating should be used in order to ensure long life.

Contactors usually have three main contacts (or poles as they are often called) and one auxiliary contact with a lower current rating. For single-phase generation the main contactor poles can be connected in parallel so that they share the current. A multiplying factor can then be applied to the three-phase current rating. For two poles in parallel the factor is 1.6 and for three poles in parallel 2.2. This allows for unbalanced current distribution between the poles.

Care should be taken to ensure that the generated voltage and frequency, under normal conditions, are suitable for the contactor coil. If the voltage is higher and/or the frequency lower than the coil rating then its life will be reduced. Reduced voltage and/or increased frequency operation is normally quite acceptable provided that the combined effect is less than 25%. For example, if the voltage is 10% below the rated coil voltage the frequency should be no more than 15% above the rated coil frequency.

If the relay contacts and contactor coil, shown connected between the load side of the auxiliary contacts and neutral, were connected directly between line and neutral from the generator the contacts would chatter (repeatedly open and close) in the event of a short-circuit or severe overload. This is because the voltage will fall and, when it is insufficient to energize the contactor coil, the contacts will open and isolate the load. With no load connected the voltage will build up, until the contactor coil is re-energized, and the contacts will close into the severe overload and cause the process to repeat itself. The repeated opening and closing of the contacts will rapidly damage them. The circuit shown in Figure A7 prevents repeated opening and closing of the contacts, once the start switch is in the run position, since if the voltage falls sufficiently to de-energise the coil the auxiliary contacts will open and prevent re-energization of the contactor coil.

The switch, S1, is initially set to the start position in order to apply voltage to the contactor coil. Provided that an overvoltage has not occurred, the d.c. relay contacts will be closed and the contactor coil will become energized. The main and auxiliary contacts will close and when the switch is set to the run position the contactor coil will remain energized. Contact chatter can occur when the switch is in the start position. This can be prevented by isolating the loads at start up with switches between the contactor and loads. However, the load must then be introduced in steps as application of full load is likely to collapse the voltage and cause the contacts to open. For three-phase systems, provided that no three-phase loads are connected, each phase can be switched separately, with the voltage being returned to rated voltage between switching operations by increasing the turbine power output. For a single phase system, unless the transmission system can be conveniently split so that there is more than one line, it will be necessary to switch full load directly with the contactor. In this case the turbine must be started rapidly to minimise contactor wear and the contactor should be overrated by a factor of at least two.

The d.c. relay and start switch should be rated for a current of at least 10 Amps, since the inrush current into the contactor is quite large and highly inductive.

The voltage sensing circuit is the same for single-phase and three-phase systems. Voltage sensing between just one line and neutral is adequate for a three-phase system, because even with considerable phase inbalance the variation between phase voltages is quite small. The sensing circuit has to be able to withstand the full overvoltage that occurs when the load is disconnected and the turbine-generator overspeeds. This should only last for a short time since the excitation capacitors will be switched out by the MCB(s), as explained in Chapter 8. However, in case the MCB(s) fail to operate, the circuit has been designed to withstand a continuous a.c. voltage of up to 600 V, which is higher than the maximum runaway voltage for all but very large generators, as explained in Chapter 8.

Power Circuit

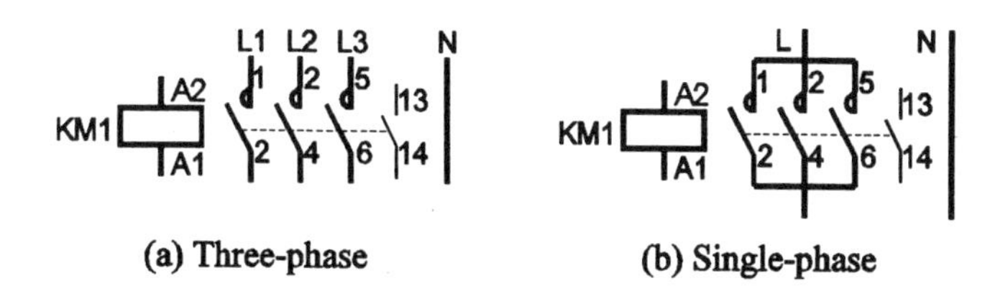

Control Circuit

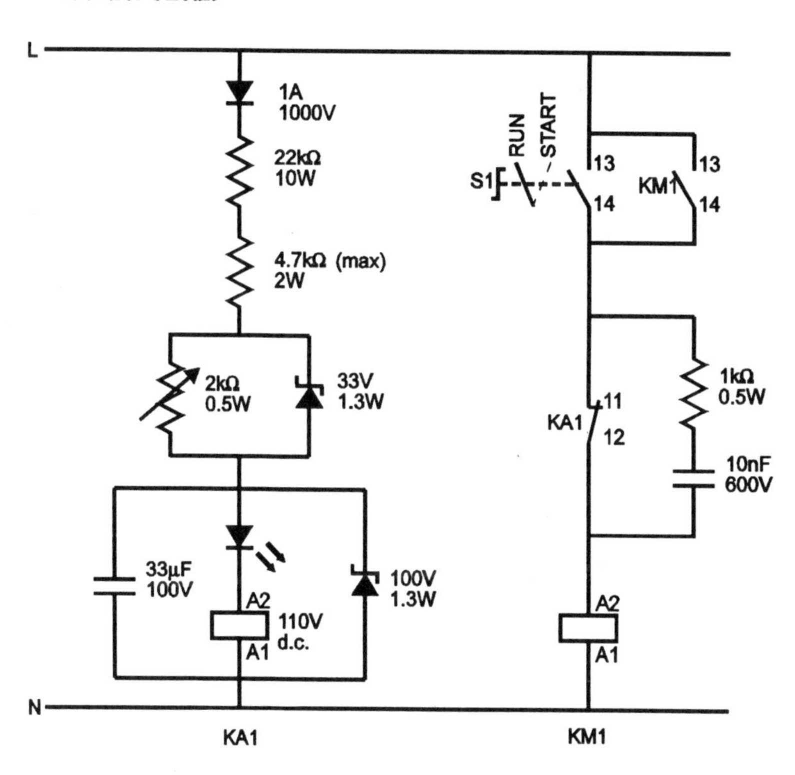

Figure A7 Circuit diagram for overvoltage trip.

The two power resistors are selected to set the operating voltage range for the trip. The 22 kΩ resistor will be used for all trips. The 2 Watt resistor can be selected with any value up to 4.7 kΩ, the higher the value the greater the voltage required for the trip to occur.

In series with the two power resistors is a variable resistor which allows some on site adjustment to be made. If no on site adjustment is required then a shorting link can be used instead of this component. The value of the variable resistor has been restricted to 2 kΩ for two reasons:

1) It prevents the trip voltage from being set too high, which could be damaging to the loads.
2) It prevents overheating of the variable resistor, since devices with a power rating above 0.5 W are rare and expensive.

Due to the tolerances of the resistors and the operating voltage of the relay, the voltage range of the trip should be determined by measurement. As a guide, with just the 22 kΩ power resistor in circuit the tripping voltage range will be approximately 230 V to 250 V, depending on the setting of the variable resistor. Higher voltages can be achieved by adding a 2 Watt resistor of suitable value.

The zener diodes are used to protect the variable resistor, capacitor and relay from being burnt out due to an overvoltage. The resistor and capacitor across the relay contacts are to reduce arcing when the contacts operate.

A relay with an a.c. coil would appear to be the obvious choice for this circuit. However, these exhibit appreciable contact chatter when the voltage just reaches the operating voltage of the coil and therefore they are not suitable. A d.c. relay has been used instead. A 110 V d.c. relay with a coil resistance of 19,000 Ω has been chosen, since lower voltage relays require a higher operating current and therefore produce greater power dissipation in the power resistor and variable resistor. Unfortunately relays with such a high coil voltage and resistance are difficult to obtain. However, a full kit of components, including the circuit board, can be purchased through ITDG.

Appendix 6: Power factor correction for motor starting

The unit described here is for single-phase motors. The circuit could be modified for three-phase motor starting, though generally star-delta starting is more appropriate, as explained in Chapter 10.

The circuit shown in Figure A8 connects the power factor correction capacitor to the supply for just 0.5 to 1.5 seconds, depending upon the setting of the variable resistor. For most applications, this is sufficient time to start the motor. If necessary the circuit values can be changed to increase the on time of the capacitor. However, if the capacitor is connected for too long the motor will become overcorrected for power factor and an overvoltage could be produced.

The capacitor should be sized to fully power factor correct the motor at the instant of starting. A large value of capacitance is required. However, due to its intermittent use, a 'motor start' type capacitor can be used. The circuit is connected after the supply switch, as shown in Figure A9.

The switching element which connects and disconnects the start capacitor is a d.c. operated solid-state relay. These devices have a higher current rating for intermittent operation than for continuous operation, as specified in the manufacturers' data sheets.

When the supply is connected the 33 μF capacitor, C1, is rapidly charged via the diode bridge and presents a d.c. source to the RC circuit to the right of it. The resistive part of this RC circuit is provided by the fixed resistor R3, the variable resistor VR1 and the resistance in the input circuit of the solid-state relay, as shown in Figure A10.

The solid-state relay (SSR) is treated as a current controlled device. When power is supplied to the circuit, by switching on the motor, a current will flow through the input side of the SSR charging C2 and causing the SSR to switch in the power factor correction capacitor. As C2 charges, the current flowing gradually falls and the SSR will turn off, disconnecting the excitation capacitor. With the component values shown, this takes approximately 0.5 seconds for VR1 equal to 0 Ω and 1.5 seconds for VR1 equal to 47 kΩ. However, much will depend upon the minimum operating voltage of the SSR. Some indication of the switching time can be obtained by connecting an analogue ammeter in series with the power factor correction capacitor. If a storage oscilloscope is available then the time delay can be measured accurately.

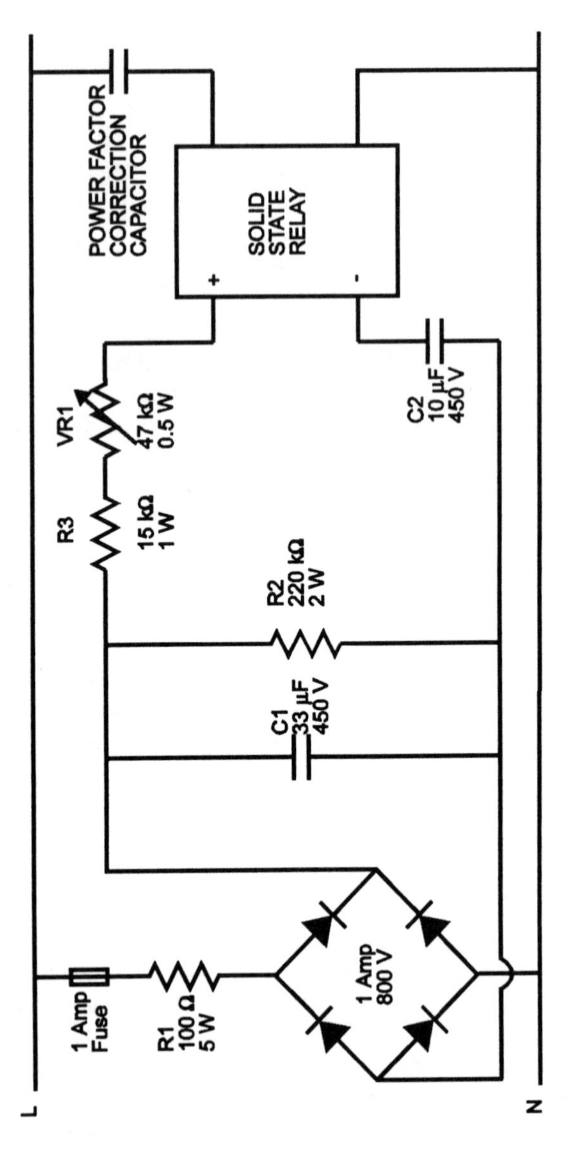

Figure A8   Temporary power factor correction for motor starting

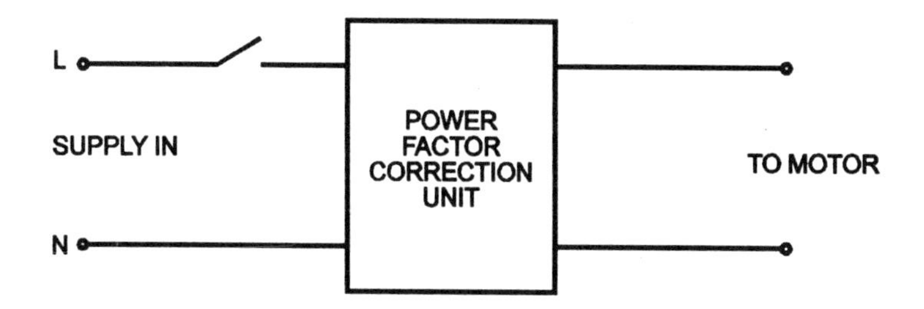

Figure A9 Connection of power factor correction unit

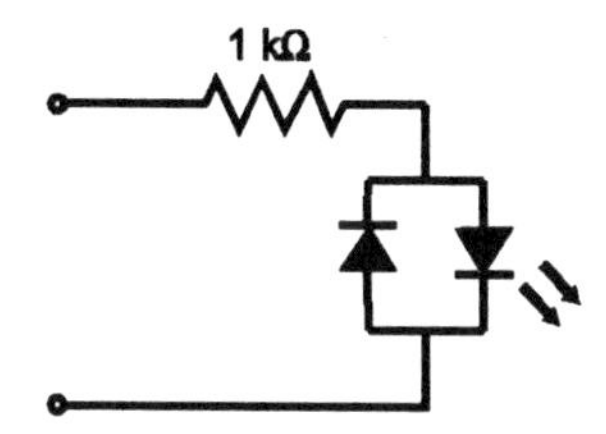

Figure A10 Typical input circuit for a d.c. operated solid-state relay

The SSR should be rated for twice the peak supply voltage, since when the capacitor is switched out it will initially remain charged and cause a peak voltage of twice the supply across the SSR output. In addition, the peak input voltage to the solid-state relay must be less than its maximum rating. The maximum peak input voltage is given by:

$$V_{PK} = \frac{\sqrt{2}V_{rms}.R_{SSR}}{(R_3+R_{SSR})}$$

Where V_{rms} is the r.m.s. supply voltage

R_{SSR} is the input resistance of the SSR

Note that when using this circuit with fridges and freezers and any other loads that contain a thermostat the circuit must be connected after this switching element in order for it to operate every time the motor is switched in.

INDEX